# The Professional's Guide To Maintenance And Reliability Terminology

# The Professional's Guide To Maintenance And Reliability Terminology

By Ramesh Gulati, Jerry Kahn and Robert Baldwin

Hard Cover ISBN 978-0-9825163-7-9
Paperback ISBN 978-0-9825163-6-2

---

© Copyright 2010 Reliabilityweb.com.

All rights reserved.

Printed in United States of America

This book, or any parts thereof, may not be reproduced, stored in a retrieval system, or transmitted in any form without the permission of the publisher.

Opinions expressed in this book are solely the authors and do not necessarily reflect the views of the Publisher.

Cover design by Patricia Serio

For information: Reliabilityweb.com
www.reliabilityweb.com
PO Box 60075 Fort Myers, FL 33906
Toll Free: 888-575-1245, Office: 239-333-2500
E-mail: customerservice@reliabilityweb.com

10  9  8  7  6  5  4  3  2  1

# Contents

Foreword ............................................................................................................ IX
Acknowledgements ........................................................................................... XI
Publisher's Foreword ...................................................................................... XIII
Preface ............................................................................................................ XV
A ........................................................................................................................ 1
B ...................................................................................................................... 21
C ...................................................................................................................... 33
D ...................................................................................................................... 67
E ...................................................................................................................... 89
F .................................................................................................................... 103
G .................................................................................................................... 119
H .................................................................................................................... 125
I ..................................................................................................................... 133
J ..................................................................................................................... 145
K .................................................................................................................... 147
L ..................................................................................................................... 151
M .................................................................................................................... 159
N .................................................................................................................... 179
O .................................................................................................................... 185
P ..................................................................................................................... 195
Q .................................................................................................................... 221
R .................................................................................................................... 227
S ..................................................................................................................... 249
T .................................................................................................................... 285
U .................................................................................................................... 303
V .................................................................................................................... 307
W .................................................................................................................... 315
X .................................................................................................................... 323
Y .................................................................................................................... 325
Z ..................................................................................................................... 327
Acronyms & Initialisms .................................................................................. 329
Symbols & Numbers ...................................................................................... 353
Resources on the Web .................................................................................... 357
Bibliography ................................................................................................... 359
Biographies .................................................................................................... 363

The authors dedicate this book to maintenance and reliability professionals around the world.

# Foreword

This document represents a significant step towards improving the knowledge of and communications between members of the Maintenance and Reliability profession. It will also increase the understanding of representatives from other professions with whom they interface. The last known significant glossary of maintenance and reliability terms, published in 1998, contained just over 1,000 entries. With more than 3,000 entries, this compilation reflects a virtual explosion of commonly practiced concepts, ideas, methodologies and various approaches to maintenance and reliability. These have improved the capacity, productivity, competitiveness, effectiveness, and global reach of manufacturing and service organizations, and commercial and government entities in recent years. This updated aggregation of terms, acronyms and definitions also is the result of a larger, more diverse cross section of contributors than ever attempted previously.

The organizers of this effort, Bob Baldwin, Ramesh Gulati and Jerry Kahn, have each served the maintenance and reliability profession in many capacities for decades. Together, they have over 100 years of experience working in this field. All are Certified Maintenance and Reliability Professionals (CMRPs).

This document will have international impact and bring about more comprehensive cooperation and improved dialog between maintenance and reliability professionals in countries around the world. Among many uses, it should become a study resource used by candidates for certification exams by bodies such as the SMRP Certifying Organization (SMRPCO). These and all others who value staying current in professional knowledge will want this dictionary readily at hand for routine referral.

As with all such compilations, the effort to update it is never completed. With the increasing number of bright, innovative entrants into the maintenance and reliability profession, many more concepts, ideas, and service provider offerings with their unique terms, abbreviations and meanings will come into common use and need to be added. This implies that future revisions and expansions will be needed to this vital collection of essential information.

Accordingly, when users find deficiencies or omissions they should call them to the attention of the publisher or the authors so that the process for corrections and additions can be initiated and new editions of this document generated.

Sincerely,

Jack Nicholas Jr., BS (Marine Engineering), MBA, P.E., CMRP

# Acknowledgements

The concept for this book began several years ago when we were working together on the Best Practices and Body of Knowledge committees of the Society of Maintenance and Reliability Professionals (SMRP). We could not find a comprehensive source for terms and definitions related to maintenance and reliability, so we decided to develop one. Also, many professional friends in this field persuaded us to take on this project.

We would like to acknowledge those who shared their valuable time as reviewers, and were a sounding board to enhance the book's content. They are Jack Nicholas, Jr., Kay Goulding of Meridiam, Dan Williams, retired senior engineer at AEDC, and Vijay Narain, from AEDC. We also thank Bart Jones and David Hurst, from AEDC, for their continued support of the project.

We acknowledge our publisher, Terrence O'Hanlon, a good friend and fellow professional, who constantly encouraged and supported development of this book.

We also acknowledge the publishing staff at Reliabilityweb.com, specifically Sylvia Hermreck and Patricia Serio, for their continued support and patience with our constantly changing schedule and text revisions.

Finally, we want to acknowledge our families, without whose support we would not have been able to finish this book.

Ramesh Gulati

Jerry Kahn

Robert Baldwin

# Publisher's Foreword

In my work as a publisher, I attend many presentations about asset management, maintenance and reliability. I also edit lots of articles bound for publication. In both cases, a near universal trait among maintenance and reliability professionals is the wide spread use of three letter acronyms (TLAs), buzzwords and technical phrases.

I often try to imagine what it must be like for people who are new to maintenance and reliability listening to one of these conference presentations or reading an article chock-a-block full of these mysteries.

Various noble attempts have been made to create glossaries, lists, translations and other guides. Many glossaries can be found online and some are pretty good, if not comprehensive.

Ramesh Gulati, Jerry Kahn and Robert Baldwin took a stand to create the most comprehensive and inclusive maintenance and reliability reference book ever published. This work has been in progress for several years and the end result is, as you are about to find out, mission accomplished.

The community owes each of these contributors a great debt of gratitude. Without a common set of definitions, sharing effective practices becomes more difficult. I cannot think of another team that would take this fairly thankless task and complete it in such a professional and valuable way.

In addition, the acronym section will be even more useful as I do not expect the use of three letter acronyms to decrease anytime soon among presenters and subject matter experts.

A work like this will last and stand the test of time. If you maintain a maintenance and reliability library, this book should be added immediately.

Terrence O'Hanlon, CMRP
Publisher
Reliabilityweb.com®
Uptime® Magazine
MRO-Zone.com

# Preface

Maintenance and reliability is not just "maintenance" plus "reliability". It is a kaleidoscope of methods, techniques, and tools which are utilized to maximize uptime. The maintenance technician applies his skills, acquired through training and experience. The maintenance manager leads his organization on a journey of continuous improvement. The reliability engineer analyzes data and explores ways to improve equipment performance. The maintenance planner and scheduler endeavors to minimize impacts on production. The external service vendor applies his or her trade to supplement the organization's core maintenance competencies.

Maintenance and reliability involves many different people in many different roles. If they are expected to work efficiently, productively and harmoniously on tasks and projects, they need a common understanding and language for communication. It is the goal of this book to provide that basis.

It has been more than a decade since the last publication of a compilation of maintenance and reliability terms, a time during which many advances have been made in technology and in the social sciences. New standards have been promulgated and new workplace regulations passed. Many new acronyms have been introduced. The World Wide Web has supplanted library stacks as a research tool. The workforce makeup has changed and the role of the maintenance and reliability employee has evolved. All these factors have affected the vocabulary needed to communicate.

This first edition of *The Professional's Guide to Maintenance and Reliability Terminology* contains over 3,000 up-to-date terms and acronyms that are encountered in the field of maintenance and reliability. Also within these pages are terms related to quality and project management, since both doctrines are required to complete tasks or implement improvement programs. There are also terms relating to sociology and communication, since these skills are required for people to work together to achieve common goals. And there are terms related to work place safety, since this must always be foremost in carrying out daily tasks.

The terms are arranged alphabetically. A complete list of acronyms and initialisms follows the main definitions. Useful web sites and a comprehensive bibliography are provided for those seeking definitions for terms that are beyond the scope of this reference.

*"The beginning of wisdom is a definition of terms"*
– Socrates

# A

| | |
|---|---|
| **A/D Converter** | A device that changes an analog (A) signal such as voltage or current into a digital (D) signal (consists of discrete data values). |
| $A_i$ | Availability (inherent). |
| $A_o$ | Availability (operational). |
| **ABC Classification** | A method of classifying items involved in a decision situation on the basis of their relative importance. Classification may be on the basis of monetary value, availability of resources, variations in lead-time, part criticality to the running of a facility and others. |
| **ABC Inventory Control** | An inventory control methodology used to optimize inventory levels and cost based on the ABC classification. |
| **Abrasion** | A general wearing, grinding, or rubbing away of a surface by friction, usually due to the presence of foreign matter such as dirt, grit, or metallic particles in the lubricant. It may also cause a breakdown of the material (such as the tooth surfaces of gears). Lack of proper lubrication may result in abrasion. |
| **Abrasive Wear** | Wear caused by abrasion. |
| **Abscissa** | The horizontal (X) axis of a chart or graph. The vertical (Y) axis is called ordinate. |
| **Absolute** | A term applied to calibration (e.g. of an accelerometer) based upon the primary standards of mass, length and time. |
| **Absolute Accuracy** | A measure of the uncertainty of an instrument reading compared to that of a primary standard. |
| **Absolute Filtration Rating** | The largest diameter of hard spherical particle that will pass through a filter under specified test conditions. This is an indicator of the largest opening in the filter elements. |

**Absolute Pressure** — Pressure measured with respect to absolute zero pressure. Distinct from pressure measured with respect to some standard pressure, e.g., atmospheric pressure, it is the sum of the available atmospheric pressure and the gage pressure.

**Absolute Temperature** — The temperature measured from absolute zero using an absolute temperature scale (e.g. Kelvin and Rankin).

**Absolute Vibration** — Vibration of an object relative to a fixed point in space. Seismic sensors, accelerometers and velocity pickups, measure absolute vibration.

**Absolute Viscosity** — A term used interchangeably with viscosity to distinguish it from either commercial or kinematic viscosity. Absolute viscosity is the ratio of shear stress to shear rate. It is a fluid's internal resistance to flow. The common unit of absolute viscosity is the poise.

**Absolute Zero** — The temperature at which all molecular motion ceases (-459.67° Fahrenheit, -273.15° Celsius, 0° Rankin, and 0° Kelvin).

**Absorbent Filter** — A filter medium that holds contaminant by mechanical means.

**Absorber** — A device capable of soaking up vibration

**Absorption** — The assimilation of one material into another. In petroleum refining, the use of an absorptive liquid to selectively remove components from a process stream.

**Absorptivity** — The proportion of the radiant energy impinging on a material's surface that is absorbed into the material. For a blackbody, this is unity (1.0).

**Accelerated Deterioration** — Deterioration of equipment that occurs over a relatively short period of time.

**Accelerated Life Testing (ALT)** — An activity during development of a new product that subjects prototypes to stress levels (including vibration, usually random) that are much higher than those anticipated in the field. The purpose is to identify failure-prone, marginally-strong elements. Those elements are strengthened and tests are continued at higher levels. Synonymous with *Test, Analyze and Fix (TAAF) testing*.

**Accelerated Stress Testing** — A post-production activity on a sampling of units. The intent is to precipitate hidden or latent failures caused by poor workmanship and to prevent flawed units from reaching the next level of assembly or the customer. Intensity is typically half that achieved in accelerated life testing.

**Accelerated Test** — A test in which the applied stress level is chosen to exceed that stated in the reference conditions in order to shorten the time required to observe the stress response of the item, or to magnify the response in a given duration. To be valid, an accelerated test must not alter the basic modes and/or mechanisms of failure.

**Acceleration** — Acceleration is rate of change of velocity with time (denoted as dv/dt), usually along a specified axis, and expressed in g or gravitational units. (1g = 32.17 ft/s2). Acceleration measurements are usually made with accelerometers.

**Accelerometer** — A sensor, transducer, or pickup that converts acceleration to an electrical signal. Two common types are piezoresistive and piezoelectric.

**Acceptable Quality Level (AQL)** — A quality level that, for the purpose of sampling inspection, is the limit of satisfactory process average.

**Acceptance Criteria** — Requirements that a project or system must demonstrably meet before customers accept delivery.

**Acceptance Number** — The maximum number of defects allowable in a sampling lot for the lot to be acceptable.

**Acceptance Sampling** — Inspection of a sample from a lot to decide whether to accept that lot. There are two types: attributes sampling and variables sampling.

**Acceptance Sampling Plan** — A specific plan that indicates the sampling sizes and associated acceptance or non-acceptance criteria to be used.

**Acceptance Test** — A test conducted under specified conditions using delivered or deliverable items in order to determine the item's compliance with specified requirements.

**Access** — To gain entry to part of a system or piece of equipment.

**Accessibility**  A measure of the related ease of admission to the various areas of an asset or system for the purpose of operation or maintenance.

**Accountability**  Answerable, but not necessarily personally charged with doing the work. Accountability cannot be delegated but it can be shared.

**Accounting Period**  A set period of time, usually a month, quarter, or year, in which costs and revenues are posted for information and analysis.

**Accounts Payable**  The amount of money owed for goods or services that were purchased on credit.

**Accreditation**  A process in which certification of competency, authority, or credibility is established. Organizations that certify third parties against official standards are themselves formally accredited by accreditation bodies (such as ANSI). They are sometimes known as "accredited certification bodies". The accreditation process ensures that their certification practices are acceptable, typically meaning that they are competent to test and certify third parties, behave ethically, and employ suitable quality assurance.

**Accumulator**  A container in which fluid energy is stored under pressure as a source of fluid power.

**Accuracy**  The characteristic of a measurement that tells how close an observed value is to a true or reference value.

**Achieved Availability**  The probability that an equipment or system, when used under stated conditions in an ideal support environment will operate satisfactorily any time when needed. This definition is similar to Ai - Inherent Availability except that it includes corrective and preventive maintenance actions but excludes logistics and other delays in downtime for calculating availability.

**Acidity**  In lubricants, acidity denotes the presence of acid-type constituents whose concentration is usually defined in terms of total acid number. The constituents vary in nature and may or may not markedly influence the behavior of the lubricant.

| | |
|---|---|
| **Acquisition** | Obtaining supplies or services by and for the use of an organization through a purchase or lease, regardless of whether the supplies or services are already in existence. |
| **Action Plan** | A specific method, process, or series of steps taken to implement the actions needed to achieve strategic goals and objectives. |
| **Active Inventory** | The raw materials, items, components, spares, work in process, and finished products that will be used or sold within a given time period. |
| **Active Listening** | Paying close attention to what is said, asking the other party to describe carefully and clearly what is meant, and requesting that ideas be repeated to clarify any ambiguity or uncertainty. |
| **Active Redundancy** | Having assets, processes, and procedures that carry out the same operations concurrently. Should one of these items fail, the system will continue to operate in a normal mode. |
| **Active Repair Time** | That portion of the downtime during which one or more maintainers are working on the equipment or system to affect a repair. This time includes preparation time, fault-correction time, and final checkout time for the system. |
| **Active Voice** | A style of writing where the object of the action is the subject of the sentence. |
| **Activity** | Element of work that is required, uses resources, and takes time to complete. Activities have expected durations, costs, and resource requirements and may be subdivided into tasks. |
| **Activity Analysis** | Identification and description of activities in an organization, and evaluation of their impact on its operations. Activity analysis determines what activities are executed, how many people perform the activities, how much time they spend on them, how much and which resources are consumed, what operational data best reflects the performance of the activities, and of what value the activities are to the organization. |
| **Activity-Based Costing** | An accounting method that assigns costs to products or services based on the amount of resources they consume. |

| | |
|---|---|
| **Activity Board** | An information-sharing display prepared by a team or group to facilitate communication between operators and maintainers. |
| **Activity Duration** | The time needed to accomplish the work involved in an activity, considering the nature of the work and resources required for it. |
| **Activity Network Diagram** | An arrow diagram used in planning and managing processes and projects. |
| **Actual Cost to Planning Estimate** | The ratio of the actual cost incurred on a work order to the planner's estimated cost for that work order. |
| **Actual Finish Date** | Point in time when work ended on an activity. |
| **Actual Operating Time** | The number of hours the equipment, asset, or system is performing its required function. |
| **Actual Production Rate** | The rate at which an asset actually produces product during a specified time period. |
| **Actual Start Date** | Point in time when work started on an activity. |
| **Actual Work Order Cost** | The cost of a work order after it is closed. It includes the cost of labor hours (mechanic, operator, electrician, etc.), service (contractors, vendors, etc.), material and service parts requirements (repair parts) and special tools/equipment (crane). |
| **Actual Work Order Hours** | The quantity of hours reported on a work order after it is closed. |
| **Actuarial Analysis** | Statistical analysis of failure data to determine the age-reliability characteristics of an item. |
| **Ad Hoc Committee** | A committee created for a short-term purpose. |
| **Additive** | A compound that enhances or imparts some property to a base material. In hydraulic fluid formulations, the more important types of additives include anti-oxidants, anti-wear additives, corrosion inhibitors, viscosity index improvers, and foam suppressants. |

| | |
|---|---|
| **Additive Stability** | The ability of additives in a fluid to resist changes in their performance during storage or use. |
| **Address – Computer** | A number specifying a location in memory where data is stored. |
| **Address – Network** | A name or token specifying a particular computer or site on the Internet or other network. |
| **Adhesion** | The property of a lubricant that causes it to cling or adhere to a solid surface. |
| **Adiabatic Process** | A process in which heat is neither lost nor gained. |
| **Adjustments** | Minor tune-up or adjusting actions requiring only hand tools, no parts, and usually lasting only a few minutes. |
| **Administrative Expense** | Expense that cannot be easily identified with a specific function or project but contributes in some way to the project or general business operations. |
| **Administrative Support** | Services such as budgeting, purchasing, information technology, and data processing which are necessary to support core operations. |
| **Administrative Time** | Time that an asset is not available to run due to a business decision (e.g., economic). Synonymous with *administrative delay* and *administrative idle time*. |
| **Adsorbent Filter** | A filter medium primarily intended to hold soluble and insoluble contaminants on its surface by molecular adhesion. |
| **Adsorption** | A process in which a gas, liquid, or solid adheres to the surface of a solid or (less frequently) a liquid but does not penetrate it, such as in adsorption of gases by activated carbon (charcoal). |
| **Adsorptive Filtration** | The attraction to, and retention of, particles in a filter medium by electrostatic forces or by molecular attraction between the particles and the medium. |
| **Aeration** | The state of air being suspended in a liquid such as a lubricant or hydraulic fluid. |
| **Aesthetics** | A dimension of quality that refers to subjective sensory characteristics such as taste, sound, appearance, and smell. |

**Affinity Diagram** — An analytical tool that helps to organize information, ideas, data, facts, opinions and issues in naturally related groups. It is especially useful for large or complex problems. Synonymous with *affinity chart* and *affinity analysis*.

**Age Exploration** — An iterative process used to optimize preventive maintenance (PM) intervals. For example, if a PM calls to replace V-belt or oil seal on an annual basis, or overhaul of an electric motor on a three year interval, its condition, as found, is measured and documented and the PM interval is increased by some amount (e.g., 2 months or by 10-20 percent). At the next PM schedule, the same procedure is repeated until an optimum PM interval is established.

**Agenda** — An outline of topics to be addressed during a meeting detailing the length of time and person(s) responsible for each topic.

**Agglomeration** — The potential of a system for particle attraction and adhesion.

**Agile Manufacturing** — A term applied to an organization that has created the processes, tools, and training in order to respond quickly to customer needs and market changes while still controlling costs and quality.

**Agreement** — Mutual assent between two or more competent parties and usually reduced to writing in a contract.

**Air Breather** — A device permitting air movement between atmosphere and the component in/on which it is installed.

**Air Changes** — The rate of air ventilation expressed as the number of times per hour that the air volume in a given space (such as a room or building) is changed by a ventilating system. It equals the cubic-feet-per-hour of air flow in and out of a specified space divided by the cubic feet of the space.

**Air Gap** — The space between magnetic poles or between rotating and stationary assemblies in a motor or generator.

**Air Vent** — A fitting used to vent air manually or automatically from a system.

**Air-to-Air Heat Exchanger** — A device in which latent heat is transferred from a warm air stream (exhaust) to a cooler air stream (incoming).

| | |
|---|---|
| **Air-to-Air Heat Pump** | A refrigeration or air conditioning system that removes heat from one air source and transfers it to another air stream. It is usually applied to heat inside air by cooling outside air. When operated in reverse, the system will perform like a typical air conditioning system by rejecting heat to the outside air. |
| **Airborne Ultrasonic** | A technology that utilizes ultrasound to locate a variety of potential problems in plants and facilities. This technology helps in leak detection, mechanical inspection of pipes and pumps, and electrical inspection. |
| **Alarm Level** | A level or condition that when reached causes an alarm to be sounded or an alert to be issued indicating that a pre-established condition of the equipment has arrived. This may require special attention or shutdown of the asset. |
| **Algorithm** | A computational procedure containing a finite sequence of steps. A set of rules that specify a sequence of actions to be taken to solve a problem. |
| **Aliasing – Vibration** | A false component in the spectrum that results when a digital sampling rate is less than two times the frequency of the data. |
| **Alignment – Equipment** | The adjustment of an object in relation to other objects. For example, aligning a motor and a compressor or pump requires one or both units be adjusted (using dial indicator or laser system) so that both shaft and coupling meet required alignment specifications (angularity and offset) to reduce vibration. A misaligned system will damage the equipment. |
| **Alignment – Organization** | Actions that are taken to ensure all process or activity support an organization's strategy, goals and objectives. |
| **Alkali** | Any substance having basic (as opposed to acidic) properties. |
| **Allocation** | The assignment or distribution for a specific purpose or to a particular person or thing. For example, the assignment or distribution of observed failures, failure rates, or reliability, to a specific organization or system for corrective action. Or the allocation of costs to different cost centers, groups, or activities. |

| | |
|---|---|
| **Alternating Current (AC)** | An electric current that flows back and forth, typically changing direction 50 or 60 times per second. |
| **Ambient Environment** | The conditions (e.g., temperature and humidity) characterizing the air or other medium that surrounds an object. |
| **Ambient Noise** | The total noise present in an environment. Usually a composite of sounds from many sources, including characteristic and background noise. |
| **Amortization** | In accounting, the process by which the cost of an intangible asset (such as an intellectual property right) is distributed over the projected useful life of the asset. |
| **Amp** | Short for ampere, the standard unit for measuring the strength of an electric current. |
| **Amplification Factor – Synchronous** | A measure of the susceptibility of a rotor to vibration amplitude when rotational speed is equal to the rotor natural frequency. |
| **Amplifier** | A device or instrument whose output is an increased function of an input signal, drawing power from a source other than from the input signal. |
| **Amplitude** | The magnitude of dynamic motion or vibration. |
| **Amplitude Modulation** | The process where the amplitude of a signal is varied as a function of the instantaneous value of another signal. The first signal is called the carrier, and the second signal is called the modulating signal. |
| **Analog** | Pertaining to data consisting of continuously variable physical quantities. |
| **Analog Signal** | A form of data transmission where the signal varies continuously in direct proportion to the intensity of the physical quantity, property, or condition represented. Analog is usually thought of in an electrical context. However, mechanical, pneumatic, hydraulic, and other systems may also convey analog signals. |
| **Analog-to-Digital Convertor** | Circuit whose input is information in analog form and whose output is the same information in digital form. |

| | |
|---|---|
| **Analysis** | A step-by-step process for determining the solution to a problem. |
| **Analysis – Chemical** | The process of determining the composition of a substance or material using chemical or physical methods. |
| **Analysis – Data** | The collection, viewing, and drawing of conclusions from data and information. |
| **ANalysis Of Means (ANOM)** | A statistical procedure for troubleshooting industrial processes and analyzing the results of experimental designs with factors at fixed levels. It provides a graphical display of data. Analysis of means is easier for quality practitioners to use because it is an extension of the control chart. |
| **ANalysis Of VAriance (ANOVA)** | A basic statistical technique for analyzing experimental data. It subdivides the total variation of a data set into meaningful component parts associated with specific sources of variation in order to test a hypothesis on the parameters of the model, or to estimate variance components. There are three models: fixed, random and mixed. |
| **Analytical Ferrography** | The magnetic precipitation and subsequent analysis of wear debris from a fluid sample. This approach involves passing a volume of fluid over a chemically treated microscope slide which is supported over a magnetic field. Permanent magnets are arranged in such a way as to create varying field strengths over the length of the substrate. This varying strength leads wear debris to precipitate in a distribution with respect to size and mass over the ferrogram. Once rinsed and fixed to the substrate, this debris deposit serves as a media for optical analysis of the composite wear particulates |
| **Analytical Thinking** | A methodical step-by-step process of breaking down a problem or situation into discrete parts to understand how each part contributes to the whole. Synonymous with *analytical approach*. |
| **AND Gate** | A logic gate for which an output occurs if all inputs coexist. All inputs are necessary and sufficient to cause the output to occur. |
| **Angular Frequency** | The torsional vibration frequency in radians per second. |

**Angularity** — In alignment, the angles between two centerlines.

**Anhydrous** — Devoid of water.

**Annealing** — Process of heating and cooling a material, usually to reduce residual stresses or to make it softer.

**Annunciator** — A device that gives audible or visible warning or alarm when a measured process variable differs from a predetermined value.

**Anodize** — To protect from corrosion by electrolytically depositing an oxide of aluminum or magnesium on a surface.

**Anti-Aliasing Filter** — Most commonly, a low-pass filter designed to filter out frequencies higher than half the sample rate in order to minimize aliasing.

**Anti-Foam Agent** — Additives used to reduce foaming in petroleum products. Silicone oil is used to break up large surface bubbles while various kinds of polymers are used to decrease the amount of small bubbles entrained in the oils.

**Anti-Friction Bearing** — A rolling contact type bearing in which the rotating or moving member is supported or guided by means of ball or roller elements.

**Anti-Oxidants** — Additives that prolong the induction period of a base oil in the presence of oxidizing conditions and catalyst metals at elevated temperatures. The additive is consumed and degradation products increase with increasing and sustained temperature as well as with increases in mechanical agitation or turbulence and contamination from air, water, metallic particles, and dust.

**Antistatic Additive** — An additive that increases the conductivity of a hydrocarbon fuel to hasten the dissipation of electrostatic charges during high-speed dispensing, thereby reducing the fire or explosion hazard.

**Antistatic Agent** — Any fault avoidance technique which reduces or eliminates the possibility of electrostatic discharge which can degrade item performance.

| | |
|---|---|
| **Antiwear Additives** | Additives that improve the service life of tribological elements operating in the boundary lubrication regime. Antiwear compounds (e.g., ZDDP and TCP) start decomposing at 90° to 100°C (at a lower temperature if water at 25 to 50 ppm is present). |
| **API Engine Service Categories** | Gasoline and diesel engine oil quality levels established jointly by API, SAE, and ASTM. Sometimes called SAE or API/SAE categories and formerly known as API Engine Service Classifications. |
| **API Gravity** | A measure of how heavy or light petroleum liquid is compared to water. If API gravity is greater than 10, the petroleum liquid is lighter and floats on water, if less than 10, it is heavier and sinks. API gravity is thus a measure of the relative density of a petroleum liquid to the density of water. |
| **Apparent Level** | The liquid level that appears in a sight gauge or is displayed on a control panel. |
| **Application Software** | A computer program designed to assist in the performance of a specific task, such as word processing, accounting, or inventory management. |
| **Appraisal Costs** | Costs incurred to inspect and test products or services to ascertain the level of quality and reliability attained. Costs associated with measuring, evaluating, or auditing products, components, and purchased materials to ensure conformance with quality standards and performance requirements. |
| **Apprentice** | A tradesperson (or craftsperson) in training. |
| **Appropriate Task** | A task that is both technically feasible and worth doing (applicable and effective). |
| **Aptitude** | The ability to develop requisite performance skills. |
| **Aptitude Test** | An actual or simulated trial to determine qualification or fitness to perform a task or job. |
| **Arc** | A luminous high temperature discharge produced when an electric current flows across a gap. |

| | |
|---|---|
| **Architecture – Network** | A structured set of protocols that implements a system's functions. Network architecture defines the functions, formats, interfaces, and protocols required for end users to exchange information. |
| **Arcing** | An electrical breakdown of a gas which produces ongoing plasma discharge resulting from a current flowing through a normally nonconductive media such as air. |
| **Area Maintenance** | Maintenance performed within a designated area in the plant by the maintenance shop located in the area. It often involves an area cross-functional team to evaluate, prioritize, plan, and schedule routine, preventive, and/ or capital work. |
| **Arrhenius Law** | This law states that the rate of chemical reaction is doubled (approximately) for every 10°C rise in temperature. |
| **Arrhenius Model** | A model used to determine the effect of temperature on reliability. |
| **Arrow Diagram** | A planning tool to diagram a sequence of events or activities (nodes) and the interconnectivity of such nodes. It is used for scheduling and, especially, for determining the critical path through nodes. |
| **Artificial Intelligence (AI)** | The concept that computers can be programmed to assume capabilities such as learning, reasoning, adaptation, and self-correction. |
| **AS9100** | An international quality management standard for the aerospace industry published by the Society of Automotive Engineers. Also published by other organizations worldwide as EN9100 in Europe and JIS Q 9100 in Japan. |
| **As-Built Drawings** | Construction drawings revised to show changes made during the construction process. |
| **ASCII** | An acronym for American Standard Code for Information Interchange. A universal file format for text files. It is a binary character code used to represent a character in a computer. It consists of 128 seven-bit codes for upper- and lower-case letters, numbers, punctuation, and special communication control characters. |

**As-Found and As-Left**  A document recording the condition of equipment prior to repair and after repair is completed. This document serves as a record of what changed due to maintenance and becomes part of the equipment's history.

**Ash – Tribology**  A measure of the amount of inorganic material in lubricating oil, determined by burning the oil and weighing the residue. Results are expressed as percent by weight.

**Asperities**  Microscopic projections on metal surfaces resulting from normal surface-finishing processes. Interference between opposing asperities, in sliding or rolling applications, is a source of friction and can lead to metal welding and scoring. Ideally, the lubricating film between two moving surfaces should be thicker than the combined height of the opposing asperities.

**Asperity**  The unevenness or roughness of a surface. Flat surfaces, even those polished to a mirror finish, are not truly flat on an atomic scale. They are uneven, with sharp, rough, or rugged outgrowths, termed asperities.

**Assembler – Computer**  A program that converts assembly language programs, which are understandable by humans, into executable machine language.

**Assessment**  A systematic process of collecting and analyzing data to determine the current, historical, or projected status of an organization or system.

**Asset Criticality**  A ranking of assets according to potential operational impact. Criteria include safety and environmental inherent risks, replacement cost, schedule, and redundancy.

**Asset Health**  The overall condition of an asset measured in terms of reliability or technical assessments from predictive testing, operational monitoring, or maintenance history analysis.

**Asset Hierarchy**  A diagramming of facility assets with parent-child relationships to the degree necessary for work management, cost and reliability analysis, and property identification.

**Asset Life Cycle**  The phases of an asset's life cycle which include: design-development, build and install, operations, maintenance, and decommissioning or disposal. Synonymous with *asset lifetime*.

| | |
|---|---|
| **Asset Management** | The set of methods, procedures, and tools to optimize the impact of costs, performance, and risk exposures (e.g., availability, efficiency, quality, longevity, and regulatory, safety and environmental compliance) of the company's physical assets. |
| **Asset Management Plan** | A documented asset description including a hierarchical breakdown, listings of interfaces and interdependencies, asset management strategy, significant histories, criticalities, replacement cost, and other specifications as necessary to configure the asset. |
| **Asset Number** | A unique alphanumerical identification of an asset on a list, often in a maintenance information database. |
| **Asset Optimization** | A comprehensive, fully integrated, strategic plan directed to gaining and sustaining the greatest lifetime value, utilization, productivity, effectiveness, value, profitability and return on investment from physical manufacturing, production, operating and infrastructure assets. |
| **Asset Owner** | The asset owner is the person, or group of people, who has been identified by management as having responsibility for the maintenance of the confidentiality, availability, and integrity of that asset. The asset owner may change during the lifecycle of the asset. The owner does not normally, or necessarily, personally own the asset. |
| **Asset Performance Management** | A set of work processes to maximize physical asset performance, mitigate risk, and optimize cost in a business enterprise. |
| **Asset Register** | A list of all the assets in a particular workplace, together with information about those assets, such as manufacturer, vendor, make, model, specifications, etc. |
| **Asset Utilization (AU)** | The percentage of time a plant is operating at designed or demonstrated production rate, with a specified quality and defined yield. |
| **Asset Value** | The purchase price of the asset plus any costs necessary to prepare the asset for use. |

| | |
|---|---|
| **Assets** | The physical resources of an organization, such as equipment, machines, mobile fleet, systems, or their parts and components, including software, that perform a specific function or provide a service. Sometimes referred to as physical assets. |
| **Assignable Cause** | A name for the source of variation in a process that is not due to chance and therefore can be identified and eliminated. |
| **Association** | The relationship between entities or data elements in a logical data model. |
| **Assumption** | A proposition that is taken for granted. |
| **Assurance** | A dimension of service quality that refers to the knowledge and courtesy of employees, and their ability to inspire trust and confidence. |
| **Atmospheric Pressure** | The pressure of air at sea level. The pressure at which the mercury barometer stands at 760 millimeters or 30 inches, equivalent to 14.7 psia. |
| **Atomic Absorption Spectroscopy** | A technique used to measure the radiation absorbed by chemically unbound atoms by analyzing the transmitted energy relative to the incident energy at each frequency. |
| **Attenuation – Sound** | The reduction of sound intensity expressed in decibels, due to either the distance from the source of noise or to a barrier or acoustically treated material. |
| **Attenuation – Thermography** | The decrease in signal magnitude during energy transmission from one point to another. This loss may be caused by absorption, reflection, scattering of energy, other material characteristics, or may be caused by an electronic or optical device, such as an attenuator. |
| **Attribute** | A piece of information that represents a single property of an entity. |
| **Audit** | The inspection and examination of a process or quality system to ensure compliance to requirements. An audit can apply to an entire organization or may be specific to a function, process, or production step. |

| | |
|---|---|
| **Audit Trail** | A record of documentation describing actions taken, decisions made, and funds expended and earned on a project. Used to reconstruct the project, after the fact, to study lessons learned and for other purposes. |
| **Auditor** | A person, or group, that examines the validity of reported information, compliance with regulations, or the meeting of technical or certification requirements. An external auditor is from an outside firm. An internal auditor is a company employee. |
| **Authoritarian Management Style** | Management approach in which the manager tells employees or team members what is expected of them, provides specific guidance on what should be done, makes his or her role within understood, schedules work, and directs employees and team members to follow standard rules and regulations. |
| **Authority** | Power, or influence, either granted to, or developed by, an individual that leads to others doing what those individuals direct. |
| **Authorize** | To give final approval. A person who can authorize something is vested with the authority to give final endorsement, which requires no further approval. |
| **Auto Spectral Density (ASD)** | The measure of acceleration per Hz of analysis bandwidth. Also called Power Spectral Density. |
| **Autocratic Management Style** | Management approach in which the manager makes all decisions and exercises tight control over the organization. This style is usually characterized by communication from the manager downward to the employees, and not vice versa. |
| **Automatic Controller** | A device or instrument that measures the value of a process variable and then makes alterations in the flow of materials, or energy, to maintain the value of the variable within an acceptable range or limit. |
| **Autonomation** | A machine design feature that allows for stoppage if an abnormality is detected during production. It is usually associated with the Toyota production system. Synonymous with *jidoka*. |

| | |
|---|---|
| **Autonomous Maintenance** | A maintenance strategy wherein machine adjustments and minor maintenance is performed by operators who are deemed to have unique knowledge about the equipment. It is a principal pillar of Total Productive Maintenance (TPM). |
| **Autonomous Work Team** | A small group of people who are empowered to manage themselves and the work they do on a day-to-day basis. The members of an autonomous work group are usually responsible for a whole process, product, or service. They not only perform the work but also design and manage it. Synonymous with *autonomous work group*. |
| **Availability – OEE Basis** | As used in the calculation of overall equipment effectiveness (OEE), the percentage of time that an asset is actually operating (uptime) compared to when it is scheduled to operate. Synonymous with *operational availability*. |
| **Availability – Reliability Basis** | As used in reliability calculations, the probability that an item or system is operating satisfactorily, at any point in time, when used under stated conditions. Expressed by the formula: $A_i = MTBF/(MTBF + MTTR)$. Synonymous with *inherent availability*. |
| **Available** | The state of being ready for use. It includes the actual running (up state) and the periods when an asset is not running for reasons other than maintenance (idle state). |
| **Available Maintenance Hours** | The number of craft hours available during a specified period. |
| **Available Time** | The time duration when an asset is in the state of being ready for use (available). It includes the actual running time (uptime) and times when an asset is not running for reasons other than maintenance (idle time). |
| **Average Life** | How long, on average, a component or equipment will last before it suffers a failure. Commonly measured by Mean Time Between Failures (MTBF). |
| **Average Outgoing Quality (AOQ)** | The expected average quality level of outgoing product for a given value of incoming product quality. |
| **Average Outgoing Quality Limit (AOQL)** | The maximum average outgoing quality, over all possible levels of incoming quality, for a given acceptance sampling plan and disposal specification. |

| | |
|---|---|
| **Average Sample Number (ASN)** | The average number of sample units inspected per lot in reaching decisions to accept or reject. |
| **Average Total Inspection (ATI)** | The average number of units inspected per lot, including all units in rejected lots (applicable when procedure calls for 100% inspection of rejected lots). |
| **Average Training Cost per Employee** | The average cost to provide formal training to employees in an organization that is directed at improving job skills. Training costs should include all employee labor, travel expenses, documents, and registration and instructor fees. |
| **Averaging** | Summing and suitably dividing several like measurements to improve accuracy or to lessen asynchronous components. |
| **Avoidable Cost** | A cost associated with an activity that would not be incurred if the activity was not performed. |
| **Axial** | The direction along the centerline of a shaft. |
| **Axial-Load Bearing** | A bearing in which the load acts in the direction of the axis of rotation. |

# B

**Babbitt** — A soft, white, non-ferrous alloy bearing material composed principally of copper, antimony, tin, and lead.

**Baby Boomers** — People born between 1943 and 1960. They are idealist in nature and usually very loyal to their organization. They feel a sense of belonging and dedication based on their history. They are motivated by power, prestige, learning opportunities and long term benefits.

**Back Pressure** — The pressure on the outlet or downstream side of a flowing system.

**Backfill** — The process of refilling a ditch or other excavation.

**Backhoe** — A shovel that digs by pulling a boom-and-stick mounted bucket toward itself.

**Backlog** — All work that is waiting to be done. It is awaiting planning, prioritization, scheduling, and execution. Usually expressed in estimated labor-hours or weeks.

**Bactericide** — An additive included in the formulations of water-mixed cutting fluids to inhibit the growth of bacteria, promoted by the presence of water, preventing odors that can result from bacterial action.

**Balance Sheet** — A financial statement that sets forth an organization's assets, liabilities, and net worth at a particular point in time.

**Balanced Scorecard** — A strategic planning and management system that aligns business activities to the vision and strategy of the organization to improve internal and external communication, and monitor organizational performance against established goals. It provides a balanced view of the organization. The balanced score card identifies four perspectives for viewing a process or organization: Financial, Internal Processes, Learning and Growth, and Customer.

| | |
|---|---|
| **Balancing** | Adjusting the distribution of mass in a rotating element to reduce vibratory forces generated by rotation. |
| **Balancing Resonance Speed** | A rotative speed that corresponds to a natural resonance frequency. |
| **Baldrige Award** | A shortened form of the term, "Malcolm Baldrige National Quality Award". |
| **Ball Bearing** | An antifriction, rolling type bearing containing rolling elements in the form of balls. |
| **Band-Pass Filter** | A filter with a single transmission band extending from lower to upper cutoff frequencies. The width of the band is normally determined by the separation of frequencies at which amplitude is attenuated by 3 db. |
| **Bandwidth** | The frequency range (usually stated in hertz or Hz) within which a measuring system can accurately measure a quantity. Generally, the higher the bandwidth, the more information that can be sent through the system in a given time. |
| **Bar Chart** | A type of graphic in which data items are shown as rectangular bars. The bars may be displayed either vertically or horizontally and may be distinguished from one another by color or by some type of shading or pattern. |
| **Barcode** | A code representing characters by sets of parallel bars of varying thickness. They are read optically by transverse scanning. |
| **Barrier** | A roadblock to completing a task or activity. |
| **Barrier – Safety** | A countermeasure against hazards caused by a flow from an energy source to a target or resource. |
| **Barrier Analysis** | An investigative technique that involves the tracing of pathways by which an item is adversely affected by a hazard, including the identification of any failed or missing countermeasures that could have prevented the undesired. |
| **Barriers-and-Aids Analysis** | A technique used to analyze the barriers and recommended corrective action for successful implementation of tasks or activities. |

| | |
|---|---|
| **Base** | A material which neutralizes acids. An oil additive containing colloidally dispersed metal carbonate, used to reduce corrosive wear. |
| **Base Metal** | The metal to be welded, soldered, or cut. |
| **Base Stock** | The base fluid, usually a refined petroleum fraction or a selected synthetic material, into which additives are blended to produce finished lubricants. |
| **Baseline Measurements** | A set of measurements (output or metrics) used to establish the current or starting level of performance of a process, function, product, organization, etc. Baseline measures are usually established before implementing improvement activities and programs. |
| **Baseline Spectrum** | A vibration spectrum taken when a machine is in good working condition (new or just overhauled). It is used as a reference for future monitoring or analysis. |
| **BASIC** | An acronym for Beginner's All-purpose Symbolic Instruction Code, a high-level programming language developed in the mid-1960s. |
| **Basic Event** | An initiating fault or failure in a fault tree that is not developed further. These events determine the resolution limits for a fault tree analysis. Synonymous with *initiator* and *leaf*. |
| **Batch Production** | A production method adopted when the required product volumes are not adequate to permit continuous production of one product on dedicated machines. Batches of a single product are run through the process at one time which results in queues awaiting the next steps in the process. Synonymous with *batch processing*. |
| **Batch-and-Queue Production** | A production method wherein several items are produced in a batch and then moved forward to the next operation before they are all actually needed there. Thus, items need to wait in a queue. Synonymous with *batch-and-push*. Contrast with continuous flow production. |

| | |
|---|---|
| **Bathtub Curve** | A graphic representation of the relationship of the life of an asset or product versus the probability failure rate. The curve contains three phases: early or infant failure (break-in), a stable failure rate during normal use, and wear out. Synonymous with *life-history curve*. |
| **Baud Rate** | A measure of the signaling speed in a digital communication system. The number of discrete conditions or signal events per second, e.g., one baud equals one bit per second in a train of binary signals. |
| **Bead** | A narrow ridge in a sheet metal workpiece or part, commonly formed for reinforcement. |
| **Bearing** | A support or guide by means of which a moving part, such as a shaft or axle, is positioned with respect to the other parts of a mechanism. |
| **Belt Sanding** | A common operation to obtain smooth surfaces. The workpiece is held against a moving, abrasive belt until the desired degree of finish is obtained. |
| **Benchmark** | A standard measurement or reference that forms the basis for comparison. This performance level is recognized as the standard of excellence for a specific process. |
| **Benchmark Measures** | A set of measurements (or metrics) used to establish goals for improvements in processes, functions, products, etc. Benchmark measures are often derived from other firms that display "best of class" achievement. |
| **Benchmarking** | A process for measuring best practice performance and comparing the results to a standard or "best of class performance" in order to identify opportunities for improvement. The comparison to best practice, often referred to as a gap analysis, leads to a prioritized array of optimizing changes directed to gaining "best practice" levels of performance. |
| **Benchmarking Gap** | The difference in performance between the benchmark for a particular activity in one organization and the level of a like or similar activity in other organizations. It is the measured performance advantage of the benchmark organization over other organizations. |

| | |
|---|---|
| **Benchmarking Study** | A formal study aimed at determining benchmarks and practices used to attain high levels of performance. |
| **Benefit-Cost Analysis** | An examination of the relationship between the monetary cost of implementing an improvement and the monetary value of the benefits achieved by the improvement, within the same time period. |
| **Benefit-Cost Ratio** | A measure of project worth in which the equivalent benefits are divided by the equivalent costs. |
| **Best Practice** | A superior method or innovative practice that contributes to the improved performance of a process or an organization, usually recognized as "best" by other peer organizations. |
| **Best Production Rate** | The rate at which an asset is designed to produce quality product during a time period, or the historical best, whichever is higher. |
| **Best-In-Class** | An organization that has outstanding performance within an industry. Synonymous with *best practice* and *best-of-breed*. |
| **Beta Rating** | The method of comparing filter performance based on efficiency. This is done using the Multi-Pass Test which counts the number of particles of a given size before and after fluid passes through a filter. |
| **Beta Test** | A test of a product or process in its intended environment. |
| **Beta-Ratio / ß-Ratio** | The ratio of the number of particles greater than a given size upstream of a filter to the number of particles greater than the same size downstream of a filter. Filters with a higher beta ratio retain more particles and have higher efficiency. |
| **Bias** | A systematic difference between the mean of the test or measurement result and a true or reference value. |
| **Bill Of Materials (BOM)** | A listing of all the subassemblies, intermediates, parts, and raw material, that go into a parent assembly showing the quantity of each required. |

**Binomial Distribution** — The possible number of times that a particular event will occur (or not occur) in a sequence of observations. The events are binary such as go/no-go, conforming/nonconforming, or pass/fail.

**Bit** — The smallest unit of information handled by a computer. One bit expresses a 1 or a 0 in a binary numeral. Short for binary digit.

**Bitmap** — A data structure in memory that represents information in the form of a collection of individual bits. A bit map is used to represent a bit image.

**Black Belt** — A team leader or facilitator that has been certified to possess the appropriate skills to implement process improvement projects utilizing the DMAIC process - Define, Measure, Analyze, Improve and Control. This process is utilized to drive up customer satisfaction and business productivity levels.

**Black Body** — An object that absorbs 100% of the radiant energy falling on it. By definition, it has an emissivity of 1.0.

**Black Oils** — Lubricants containing asphaltic materials, which impart extra adhesiveness. They are used for open gears and steel cables.

**Blade Passing Frequency** — A potential vibration frequency on any bladed machine (e.g., turbine, axial compressor, fan, etc.) It is represented by the number of blades multiplied by the shaft-rotating frequency.

**Blanket Work Order** — A work order that is typically written to cover multiple tasks for a stated period of time or a specific area (e.g., Repair Water Leaks in the Month of July or Repair Insulation). Also used for non- specific work to allow the collection of work hours for time accounting. Synonymous with *standing work order*.

**Blemish** — An imperfection severe enough to be noticed but that should cause no real impairment with respect to intended, normal, or reasonably foreseeable use.

| | |
|---|---|
| **Block Diagram** | A diagram that shows the operation, interrelationships and interdependencies of components in a system. Boxes, or blocks, represent the components. Connecting lines between the blocks represent interfaces.<br><br>There are two types of block diagrams:<br>1. Functional Block Diagram – Shows a system's subsystems, lower level components, their interrelationships, and which interface with other systems.<br>2. Reliability Block Diagram – Similar to the functional block diagram except that it is modified to emphasize those aspects influencing reliability. |
| **Blue-Collar Worker** | An employee whose job is performed in work clothes and often involves manual labor. |
| **Blueprint** | In engineering, a line drawing showing the physical characteristics of a part, equipment, or system. |
| **Board of Directors** | A body of elected or appointed members who jointly oversee the activities of a company or organization. |
| **Bode** | Rectangular coordinates plot of 1x component amplitude and phase versus running speed. |
| **Bode Plot** | A graph of the logarithm of the transfer function of a linear, time-invariant system versus frequency, plotted with a log-frequency axis to show the system's frequency response. |
| **Body Of Knowledge (BOK)** | The prescribed aggregation of knowledge, in a particular area, an individual is expected to have mastered to be certified as, or considered, a practitioner. |
| **Boiler and Pressure Vessel Code (BPVC)** | A set of rules and guidelines governing the design, fabrication, and inspection of boilers and pressure vessels to ensure safe operation. |
| **Boiler Plate** | Standard and essential terminology and clauses found in contracts, procedures, and templates that are not subject to frequent change. |

**Boolean Algebra**  Algebra for determining whether logical propositions are true or false. In Boolean algebra, variables must have one of only two possible values, true or false, and relationships between these variables are expressed with logical operators such as AND, OR, and NOT.

**Boring**  The enlargement or relocation of an existing hole, or section of a hole, by means of a cutting tool.

**Boroscope**  Industrial endoscope, a device for viewing the interior of objects.

**Bottleneck**  A process constraint that determines the capacity or capability of a system.

**Bottom Line**  The essential or salient point. The primary or most important consideration. Also, the line at the bottom of a financial report that shows the net profit or loss.

**Boundary Lubrication**  A form of lubrication between two rubbing surfaces without development of a full-fluid lubricating film. Boundary lubrication can be made more effective by including additives in the lubricating oil that provide a stronger oil film, thus preventing excessive friction and possible scoring.

**Bow**  A shaft condition, in rotating machinery, in which the shaft centerline is not straight.

**Box Plot**  A method of graphically depicting groups of numerical data through their five-number summaries: the smallest observation (sample minimum), lower quartile (Q1), median (Q2), upper quartile (Q3), and largest observation (sample maximum). A box plot may also indicate which observations, if any, might be considered outliers.

**Brainstorming**  A problem-solving technique used by teams to generate as many ideas as possible related to a particular subject. Each person in the team is asked to think creatively and write down as many ideas as possible. The ideas are not discussed or reviewed until after the brainstorming session, at which time they are summarized.

**Brass Tag**  An equipment or device identification number.

**Break Horsepower (BHP)** — The actual or useful horsepower of a motor or engine, usually determined from the force exerted on a friction brake or dynamometer connected to the drive shaft. In pumping, it is the amount of real power going to the pump as measured at the pump inlet shaft.

**Breakdown – Equipment** — Failure of equipment, resulting in physical damage, which requires that equipment, or one of its component parts, be repaired or replaced.

**Breakdown – Project** — Identification of the smallest activities or tasks in a project for estimating, monitoring, and controlling purposes.

**Breakdown Maintenance** — Repairs or replacements performed after an asset/machine has failed to return to its functional state following a malfunction or shutdown. Synonymous with *emergency maintenance*.

**Break-Even Analysis** — Analysis used to determine the sales volume, or revenue, at which an organization is able to cover all its costs without making or losing money. Synonymous with *cost-volume profit analysis*.

**Break-In Work** — Work that disrupts, or breaks in to, the production schedule.

**Breakthrough Improvement** — A dynamic, decisive movement to a new, higher level of performance.

**Bridging** — A condition of filter element loading in which contaminant spans the space between adjacent sections of a filter element thus blocking a portion of the useful filtration.

**Bright Stock – Oil** — A heavy residual lubricant stock with low pour point used in finished blends to provide good bearing film strength, prevent scuffing, and reduce oil consumption. Usually identified by its viscosity, SUS at 210°F or cSt at 100°C.

**Brinelling** — The permanent deformation of the bearing surfaces where the rollers (or balls) contact the races. Brinelling results from excessive load or impact on stationary bearings. It is a form of mechanical damage in which metal is displaced or upset without attrition.

**British Thermal Unit (Btu)** — A unit of measure for energy or work. The heat required to raise the temperature of 1 pound of water 1° Fahrenheit.

| | |
|---|---|
| **Broadband – Communications** | Of, or relating to, communications systems in which the medium of transmission (such as a wire or fiber-optic cable) carries multiple messages at a time, each message modulated on its own carrier frequency by means of modems. |
| **Broadband – Frequency** | Vibration (or other) signals which are unfiltered. Signals at all frequencies contribute to the measured value. |
| **Broadband Network** | A local area network on which transmissions travel as radio-frequency signals over separate inbound and outbound channels. |
| **Broadband Trending** | A technique that captures the total vibration of the machine (broadband) at the specific measurement point s on the machine. The data is compared and trended, either to a baseline reading or to vibration severity charts, to determine the relative condition of the machine. |
| **Brookfield Viscosity** | The apparent viscosity in cP determined by a Brookfield viscometer, which measures the torque required to rotate a spindle at constant speed in oil of a given temperature. Basis for ASTM Method D 2983 - used for measuring low temperature viscosity of lubricants. |
| **Bubble Point** | The differential gas pressure at which the first steady stream of gas bubbles is emitted from a wetted filter element under specified test conditions. |
| **Bubble Point Test** | A procedure for measuring the largest pore, or hole, in a filter or similar fluid-permeable porous structure. |
| **Budget** | A plan that quantifies the organization's goals in terms of specific financial and operating objectives. |
| **Budget Review Process** | The process of evaluating budget proposals and arriving at the master budget. |
| **Budget Variance** | The difference between budgeted data and actual results. |
| **Buffer** | A temporary storage in front of, or following, a process or work station. |

| | |
|---|---|
| **Built-In Test (BIT)** | An integral capability test of the equipment that provides an on-board, automated testing capability, consisting of software and/or hardware components that detect, diagnose, or isolate product or system failures. |
| **Built-In Test Equipment (BITE)** | Diagnostic and checkout devices that are integrated into the equipment design so that tests can be conducted during operation. |
| **Bulk Modulus** | A measure of a substance's resistance to uniform compression. It is defined as the pressure increase needed to cause a given relative decrease in volume. |
| **Bulk Storage** | The area within a facility or warehouse devoted to the placement of large quantities of items, or single items too large to be placed in a bin storage location. |
| **Bureaucratic Elasticity** | The characteristic in which an organization begins a new initiative and then because of departure, shift of attention, lack of firm leadership of the initiating manager, or a change in management, returns to the "traditional" way of doing what the new initiative was supposed to change. |
| **Burn-In** | The process of operating devices or equipment often under accelerated voltage, temperature, or load in order to screen out infant mortality failures. |
| **Burnishing** | Plastic smearing such as may occur on metallic surfaces during buffing. |
| **Burst Pressure Rating – Filter** | The maximum specified inside-out differential pressure that can be applied to a filter element without outward structural or filter-medium failure. |
| **Bus** | A conductor, or group of conductors considered as a single entity, which transfers signals or power between elements. |
| **Busbar** | Thick strips of copper or aluminum that conduct electricity within a switchboard, distribution board, substation, or other electrical apparatus. |
| **Bushing** | A short, externally threaded connector with a smaller size internal thread. |

| | |
|---|---|
| **Business Benefits** | Benefits for the business which can be derived from cost savings (e.g., labor savings, increases in productivity, lower inventories) or from other means (e.g., enhanced market position, secured intellectual property). |
| **Business Drivers** | People, knowledge, and conditions (such as market forces) that initiate and support activities for which the business was designed. |
| **Business Process** | The end-to-end sequence of activities that defines one or more business functions required to deliver goods or services to a customer. Processes serving external customers are considered core business processes. Those serving internal customers are considered support services. |
| **Business Process Modeling (BPM)** | A method of diagramming the processes of an enterprise or organization with flow charts so that the process may be analyzed and improved. |
| **Business Process Modeling Notation (BPMN)** | The standard graphical notation used in Business Process Modeling. |
| **Business Process Reengineering (BPR)** | A method to improve organizational performance by evaluating and redesigning business processes. |
| **Business Unit** | An organization that is responsible for both the production and marketing of a product or family of related products. Synonymous with *profit center*. |
| **Bypass Filtration** | A system of filtration in which only a portion of the total flow, of a circulating fluid system, passes through a filter at any instant or in which a filter, having its own circulating pump, operates in parallel to the main flow. |
| **Bypass Valve** | A valve whose primary function is to provide an alternate flow path. |
| **Byte** | A unit of data. A byte can represent a single character, such as a letter, a digit, or a punctuation mark. |

# C

$C_{mk}$ — Machine Capability Index.

$C_p$ — Capability Index.

$C_{pk}$ — Process Capability Index.

**C+, C++** — Programming languages used in the 1990s to program many personal computer and UNIX-based applications.

**C Chart** — A control chart for evaluating the stability of a process in terms of the count of events of a given classification occurring in a sample. The average number of defects within each product for sample subgroups of equal sizes.

**Cable** — A conductor with insulation, or a stranded conductor with or without insulation and other coverings

(single-conductor cable), or a combination of conductors insulated from one another (multiple-conductor cable).

**Cache – Memory** — A special memory subsystem in which frequently used data values are duplicated for quick access. A memory cache stores the contents of frequently accessed RAM locations and the addresses where these data items are stored.

**CAD / CAM** — An acronym for the integration of computer-aided design (CAD) and computer-aided manufacturing (CAM) to achieve automation for design and manufacturing.

**Cage – Bearing** — The separator that spaces and holds rolling elements in their proper positions along the races.

**Calibration** — A comparison of a measurement of a system of unverified accuracy to a measurement of a system of known accuracy in order to detect any variation from true value. In a broader sense, it is the process of determining and/or adjusting an entity or device to meet or match a predetermined set of conditions or standards.

| | |
|---|---|
| **Call Back** | A job where the maintenance person is called back because the asset broke again or the job wasn't finished the first time. |
| **Call-Out** | To summon a tradesperson to the workplace during his non-working time so that he can perform a maintenance activity, usually an emergency maintenance task. |
| **Campbell Diagram** | A mathematically-constructed diagram used to check for coincidences of vibration sources (1′, 2′, etc., shaft speed) with rotor natural frequencies, which result in rotor resonances. It plots frequency vs. RPM. Plot size grows with increasing amplitude. Synonymous with *interference diagram*. |
| **Cannibalize** | The act of removing serviceable parts from one item of equipment to install them on another item of equipment in order to restore the latter to a serviceable condition. |
| **Capability** | The ability of an item to meet a service demand of given quantitative characteristics under given conditions, or the performance of a process demonstrated to be in a state of statistical control. |
| **Capability Analysis** | The statistical comparison of the actual performance of a process with its specification limits. "Capable" systems perform completely within specification limits as established by customer requirements. |
| **Capability Index (Cp)** | An index that measures the performance of a process with a normal distribution. Synonymous with *Process Capability*. |
| **Capability Maturity Model (CMM)** | A framework that describes the key elements of an effective software process. It's an evolutionary improvement path from an immature process to a mature, disciplined process. The CMM covers practices for planning, engineering, managing software development and maintenance. When followed, these key practices improve the ability of organizations to meet goals for cost, schedule, functionality and product quality. |
| **Capacitance** | The measure of electrical storage potential of a capacitor. The unit of capacitance is the farad. |

**Capacitance-to-Ground** — A characteristic measure of the dielectric strength of an electrical component. It is indicative of the amount of contamination (e.g., dirt or moisture) present. A component of motor circuit analysis testing.

**Capacitor** — A device that stores electrical energy.

**Capacity – Filter** — The amount of contaminants a filter will hold before an excessive pressure drop is caused. Most filters have bypass valves which open when a filter reaches its rated capacity.

**Capacity – Heat** — Ability of a material or structure to store heat. A product of the specific heat and the density of the material.

**Capacity – Production** — The highest sustainable output rate that can be achieved with the current product specifications, product mix, worker effort, equipment, and facilities.

**Capacity Factor** — The ratio of actual output divided by rated output. In the electric power generating industry, it is the ratio of actual megawatt hours produced on an annual basis to the megawatt hours that could have been produced during that period if the plant was operating at 100% of its rated output every hour of the year.

**Capillarity** — A property of a solid-liquid system manifested by the tendency of the liquid in contact with the solid to rise above or fall below the level of the surrounding liquid. This phenomenon is seen in a small bore (capillary) tube.

**Capital Asset** — A physical asset that is held by an organization for its production potential.

**Capital Budget** — A financial plan of long-term investments for the financing and acquisition of assets.

**Capital Budgeting** — The systematic process of identifying and evaluating capital investment projects to arrive at a capital expenditure budget.

**Capital Project** — Projects that include new construction, major repairs, or improvements where the cost is capitalized (the asset is then depreciated) rather than expensed.

| | |
|---|---|
| **Capital Spares** | Spares (usually large, expensive, difficult to obtain, or have long lead-times) that are acquired as part of the capital purchase of the assets. They are protection against large downtime. Accounting often treats these spares as capital items with their value depreciated over time. |
| **Capitalized Work** | Work performed on capital projects. |
| **Carbon Residue** | Coked material remaining after oil has been exposed to high temperatures under controlled conditions. |
| **Carbon Steel** | Steel that owes its properties chiefly to the presence of carbon without substantial amounts of other elements. |
| **Carbonyl Iron Powder** | A lubrication contaminant which consists of up to 99.5% pure iron spheres. |
| **Case Drain Filter** | A filter located in a line conducting fluid from a pump or motor housing to a reservoir. |
| **Case Study** | An analytical method involving an in-depth examination of a single problem, instance, or event (a case). It provides a systematic way of looking at events, collecting data, analyzing information, and reporting the results. The researcher may gain a sharpened understanding of why the instance happened as it did, and what might become important to look at more extensively in future research. |
| **Cash Flow** | The sum of all cash receipts, expenses, and investments in a project or operation at a specified point in time. |
| **Catalog – Parts** | A set of descriptions of parts, or other stock or non-stock items, which are used in the maintenance of equipment. |
| **Catalyst** | A substance which speeds a chemical action without undergoing a chemical change itself during the process. |
| **Catastrophic Failure** | An unexpected failure of an asset/machine resulting in considerable cost and downtime. |
| **Category – Work** | The types of work which make up the work load performed by maintenance personnel (preventive maintenance, emergency repairs, etc.). |

| | |
|---|---|
| **Cathode Ray Tube (CRT)** | An electron-beam tube in which the beam can be focused to a small cross-section on a luminescent screen and varied in both position and intensity to produce a visible pattern. |
| **Cause** | An identified reason for the presence of a defect or problem. |
| **Cause Mapping®** | A simple and effective method of analyzing, documenting, communicating, and solving a problem to show how individual cause-and-effect relationships are interconnected. |
| **Cause-and-Effect Analysis** | A graphical technique that can be used to identify and arrange the causes of an event, problem, or outcome. The hierarchical relationships between the causes, according to their level of importance or detail, are illustrated on branches. Synonymous with *fishbone* and *Ishikawa analysis*. |
| **Cause-and-Effect Diagram** | The diagram resulting from a cause-and-effect analysis. Synonymous with *fishbone* and *Ishikawa diagram*. |
| **Cavitation** | The formation of an air or vapor pocket (or bubble) due to lowering of pressure in a liquid. Often the result of a solid body, such as a propeller or piston, moving through the liquid. Cavitation can occur in a hydraulic system as a result of low fluid levels drawing air into the system. This produces tiny bubbles that expand explosively at the pump outlet causing metal erosion and eventual pump destruction. |
| **Cavitation Erosion** | A material-damaging process which occurs as a result of vaporous cavitations. Damage results from the hammering action when cavitation bubbles implode in the flow stream. Ultra-high pressures caused by the collapse of the vapor bubbles produce deformation, material failure, and, finally, erosion of the surfaces. |
| **Cavity Radiator** | A hole, crack, scratch, or cavity which will have a higher emissivity than the surrounding surface because reflectivity is reduced. |
| **CD-ROM** | An acronym for compact disc read-only memory. A form of storage characterized by high capacity (roughly 650 megabytes) and the use of laser optics rather than magnetic means for reading data. |

| | |
|---|---|
| **Cell** | An arrangement of people, machines, materials, and equipment with the processing steps placed right next to each other in sequential order through which parts are processed in a continuous flow. |
| **Cellular Manufacturing** | A manufacturing approach in which equipment and workstations are arranged to facilitate small-lot, continuous-flow production. In a manufacturing "cell," all operations necessary to produce a component or subassembly are performed in close proximity, thus allowing for quick feedback between operators when quality problems and other issues arise. |
| **CenterLine (CL)** | A line on a graph that represents the overall average (mean) operating level of the process. |
| **Centipoise (cp)** | A unit of absolute viscosity. 1 centipoise = 0.01 poise. |
| **Centistoke (cSt)** | A unit of kinematic viscosity. 1 centistoke = 0.01 stoke |
| **Central Limit Theorem** | The theorem that irrespective of the shape of the distribution of a population, the distribution of sample means is approximately normal when the sample size is large. |
| **Central Processing Unit (CPU)** | The computational and control unit (CPU) of a computer. The CPU is the device that interprets and executes instructions. |
| **Central Tendency** | The propensity of data collected on a process to concentrate around a value situated approximately midway between the lowest and highest values. Three common measures of central tendency include (arithmetic) mean, median, and mode. |
| **Centralized Lubrication** | A system of lubrication in which a metered amount of lubricant, or lubricants, for the bearing surfaces of a machine, or group of machines, is supplied from a central location. |
| **Centralized Maintenance** | A maintenance organization wherein a single maintenance department is responsible for the entire facility, reporting at the plant level. |

| | |
|---|---|
| **Centrifugal Fan** | A device that draws air in axially and discharges it radially. |
| **Centrifugal Pump** | A pump that uses centrifugal force to move water or other liquids. Centrifugal pumps use an impeller and a volute to create the partial vacuum and discharge pressure necessary to move water through the casing. |
| **Centrifugal Separator** | A separator that removes immiscible fluid and solid contaminants, that have a different specific gravity than the fluid being purified, by accelerating the fluid mechanically in a circular path and using the radial acceleration component to isolate these contaminants. |
| **Certification** | The result of meeting the established criteria set by an accrediting or certificate-granting organization. |
| **Certified Maintenance and Reliability Professional (CMRP)** | A certification awarded by the Society of Maintenance & Reliability Professionals Certifying Organization (SMRPCO) to an individual who demonstrates knowledge of the SMRP Body of Knowledge after successfully passing a written test. SMRPCO is an ANSI accredited organization. |
| **Certified Reliability Engineer (CRE)** | A certification awarded by the American Society of Quality to an engineer, who demonstrates an understanding of the principles of performance evaluation and prediction to improve products and systems reliability, after successfully passing a written test. |
| **Chaku-Chaku** | A Japanese word that means "load-load." It is a method of conducting single-piece flow in which the operator proceeds from machine to machine, taking a part from the previous operation and loading it in the next machine, then taking the part just removed from that machine and loading it in the following machine. Chaku-chaku lines allow different parts of a production process to be completed by one operator, eliminating the need to move around large batches of work-in-progress inventory. |
| **Champion** | A business leader, or senior manager, who ensures that resources are available for training and projects and who is involved in project tollgate reviews. An executive who supports and addresses organizational issues. |

**Change Agent** — An individual from within, or outside, an organization who facilitates change within the organization. May or may not be the initiator of the change effort.

**Change Management** — The process of bringing planned change to an organization. Change management usually means leading an organization through a series of steps to meet a defined goal. Synonymous with *management of change (MOC)*.

**Change Request** — A document that:
- contains a request for making a change to the form, fit, or function of an item or asset, or
- contains a call for an adjustment to a system or process

**Change-Out** — Remove a component or part and replace it with a new or rebuilt one.

**Changeover Time** — The time required to modify a system or workstation, usually including both tear-down time for the existing condition and set-up time for the new condition. Typically associated with a switch to new product.

**Characteristic** — The factors, elements, or measures that define and differentiate a process, function, product, service, or other entity.

**Charge Amplifier** — A device used to convert accelerometer output impedance from high to low, making calibration much less dependent on cable capacitance.

**Charge-Back** — Maintenance costs charged to the user department that requested the work.

**Chart** — A tool for organizing, summarizing and depicting data in graphic form.

**Chart of Accounts** — A numbering system used to identify and monitor project costs by category (labor, supplies, materials, etc.).

**Charter** — A documented statement officially initiating the formation of a committee, team, project, or other effort in which a clearly stated purpose and approval are conferred.

**Check Sheet** — A simple data recording device. The check sheet is custom designed by the user, allowing him or her to readily interpret the results.

| | |
|---|---|
| **Check Valve** | A valve that permits fluids to pass in one direction but closes when the fluids attempt to pass in the opposite direction. |
| **Checklist** | A tool that is used for processes where there is a large human element involved. Check lists are used as reminders or to ensure that all important steps or actions in an operation have been taken. Checklists contain items that are important or relevant to an issue or situation. |
| **Checkout** | Tests or observations of an item to determine its condition or status. |
| **Chemical Monitoring** | Monitoring that detects potential failures that cause traceable quantities of chemical elements to be released into the environment. |
| **Chemical Stability** | The tendency of a substance or mixture to resist chemical change. |
| **Chi Square Test** | Compares actual data to expected results. The test verifies or rejects a null hypothesis, which assumes no significant difference between the actual and expected data. |
| **Chip Control Filter** | A filter intended to prevent only large particles from entering a component immediately downstream. |
| **Circadian Rhythm** | A natural biological cycle, lasting approximately 24 hours, which governs sleeping and waking patterns. |
| **Circuit** | A conductor or system of conductors, through which an electric current is intended to flow. |
| **Circuit Board** | A flat board that holds chips and other components on its top side and has printed, electrically conductive paths, in multiple layers, for those components on its bottom side. |
| **Circuit Breaker** | A device that opens an electric circuit when an overload occurs. |
| **Circuit Protection** | A form of fault avoidance in which an electrical or electronic circuit is isolated from an adverse signal stress, or combination of stresses, by using conservative design approaches. |

| | |
|---|---|
| **Circulating Lubrication** | A system of lubrication in which the lubricant, after having passed through a bearing or group of bearings, is recirculated by means of a pump. |
| **Classification of Defects** | The listing of possible defects of a unit, classified according to their seriousness. Commonly used classifications:<br>• Class A, B, C, D<br>• Critical, Major, Minor, Incidental<br>• Critical, Major, Minor<br><br>Definitions of these classifications require careful preparation and tailoring to the product(s) being sampled to enable accurate assignment of a defect to the proper classification. |
| **Clean Room** | A facility or enclosure in which air content and other conditions (such as temperature, humidity, and pressure) are controlled and maintained at a specific level by special facilities, operating processes and trained personnel. |
| **Cleanable – Filter** | A filter element which, when loaded, can be restored, by a suitable process, to an acceptable percentage of its original dirt capacity. |
| **Cleaning** | Removing all sources of dirt, debris, and contamination for the purpose of inspection and locating problems. A clean and organized asset area will improve the asset's reliability and maintainability. |
| **Cleanliness Level – Lubricant** | A measure of relative freedom from contaminants. |
| **Clearance – Between Objects** | The clear distance between two objects measured surface to surface. |
| **Clearance Bearing** | A journal bearing in which the radius of the bearing surface is greater than the radius of the journal surface. |
| **Clearing Equipment or Systems** | Tagging, or locking out, the necessary valves or devices surrounding equipment to be worked on so that the equipment is safe. May also involve draining or otherwise making equipment ready to be maintained. |
| **Client** | The customer, owner, distributor, buyer, or end user of a product or service. |

| | |
|---|---|
| **Client – Computer** | A process, such as a program or task, which requests a service provided by another program. |
| **Clip Art** | A collection of proprietary or public-domain photographs, diagrams, maps, drawings, and other such graphics that can be "clipped" from the collection and incorporated into documents. |
| **Closed Loop Control** | Responses are measured and fed back to the control system to refine or modify drive signals in order to bring responses closer to the reference or desired motions. |
| **Closed-Loop Corrective Action (CLCA)** | A sophisticated engineering system designed to document, verify, and diagnose failures, recommend and initiate corrective action, provide follow-up, and maintain comprehensive statistical records. |
| **Cloud Point** | The temperature at which waxy crystals in an oil or fuel form a cloudy appearance. |
| **Coaching** | A method of directing, instructing and training a person or group of people, with the aim to achieve some goal or develop specific skills. |
| **Coalescor** | A separator that divides a mixture or emulsion of two immiscible liquids, using the interfacial tension and difference in wetting between the two liquids, on a particular porous medium. |
| **Coating** | A material that is applied in a liquid or gel state and allowed to cure to a solid protective finish. |
| **Coaxial Cable** | A physical network medium that offers large bandwidth, the ability to support high data rates with high immunity to electrical interference and a low incidence of errors. |
| **COBOL** | A compiled programming language developed in 1960 and still in widespread use today, especially in business applications. Acronym for *Common Business-Oriented Language*. |
| **Code** | Symbolic designation, used for identification. |

**Code – Computer** — Program instructions. Source code consists of human-readable statements written by a programmer in a programming language. Machine code consists of numerical instructions, that the computer can recognize and execute, that were converted from the source code.

**Code of Conduct** — Expectations of behavior mutually agreed on by an individual or team.

**Code of Ethics** — Written statement of principles addressing the behavior of the individuals employed in an organization.

**Coefficient of Friction** — The ratio of the frictional to the nominal force.

**Coefficient of Thermal Expansion** — Linear expansion or contraction per unit length per degree of temperature change between specified lower and upper temperature limits.

**Coefficient of Variation** — The ratio of the standard deviation to the mean. Since the standard deviation and the mean of a data set have the same units, the coefficient of variation (denoted by CV) will be a unit-less measure.

**Coherence – Vibration** — A measure of how much of the output signal is dependent on the input signal in a linear and time-invariant way. It is an effective means of determining the similarity of vibration at two locations, giving insight into the possibility of cause and effect relationships.

**Cohesion** — The property of a substance that causes it to resist being pulled apart by mechanical means.

**Cold Crack** — A crack that occurs in a casting, after solidification, due to excessive stress generally resulting from non-uniform cooling.

**Cold Cranking Simulator** — An intermediate shear rate viscometer that predicts the ability of oil to permit a satisfactory cranking speed to be developed in a cold engine.

**Cold Flow** — Permanent material deformation due to mechanical force or pressure but not to heat softening.

**Cold Soak** — Long term environmental exposure to lower specified temperatures.

| | |
|---|---|
| Collapse | The minimum differential pressure that an element is designed to withstand without permanent deformation. |
| Combination Maintenance Organization | A maintenance organizational structure in which the best characteristics of both centralized and decentralized maintenance are considered. In this structure, areas will have a dedicated small staff to take care of daily and routine issues, and a centralized staff responsible for major preventive maintenance (PM) activities and specialized repairs. Synonymous with *hybrid maintenance organization*. |
| Combustible | Any material that, in the form in which it is used and under the conditions anticipated, will ignite and burn or will add appreciable heat to an ambient fire. |
| Commercial Off The Shelf (COTS) | A commercially available item, such as a software product, which is purchased and used without tailoring. |
| Commissioning | A process by which equipment, a facility, or a plant (which is installed, complete, or near completion) is tested to verify that it functions in accordance with its design objectives or specifications. |
| Common Cause Event | An event which, upon occurring, induces the occurrence of two or more faults within the system. |
| Common Cause Failure Group | An event or mechanism that can cause two or more failures simultaneously is called a common cause. The failures are referred to as common cause failures. |
| Common Causes | The causes of variation that are inherent in a process over time. |
| Common Mode Failure | Failure that has the capability to bridge and defeat the redundancy factor, causing system failure by simultaneously or sequentially impacting all redundant elements. |
| Commonly Used Parts | A combination of standard replacement parts and components that may be used on a variety of equipment and systems. |
| Communication | Effective transfer of information from one party to another. Exchanging information between individuals through a common system of symbols, signs or behavior. |

| | |
|---|---|
| **Communication Barrier** | Impediment to effective communication. Barriers may be physical, environmental, cultural, temporal, psychological, emotional, linguistical, or from any other source that diminishes the transmission and receipt of a message. |
| **Community of Practice (CoP)** | An affinity group or information network that provides a forum where members can exchange tips and generate ideas. |
| **Commutator** | The part of a DC motor armature that causes the electrical current to be switched to various armature windings. |
| **Company Culture** | A system of values, beliefs, and behaviors inherent in a company. To optimize business performance, top management must define and create the necessary culture. |
| **Comparison** | A term applied to calibration (e.g. of an accelerometer) in which sensitivity is tested against a standard. |
| **Compatibility** | The ability of two devices to communicate with each other understandably or the ability of software to run on a particular hardware platform. |
| **Competency** | Critical skill, or personality characteristic, required of an individual to complete an activity or carry-out the requirements of a position. |
| **Compiler** | A program that translates all the source code of a program written in a high-level language into object code prior to execution of the program. |
| **Complacency** | A condition preventing effective decision making in which individuals either do not see, or ignore, the signs of danger or opportunity. |
| **Complainers** | A category of difficult people who continually gripe but never offer solutions. |
| **Completion Date** | The date that a work order is certified complete and closed out. |
| **Compliance** | The state of meeting requirements, which may be prescribed specifications, contract terms, metrics, regulations, or standards. |

**Component**  An item or subassembly of an asset, usually modular and replaceable, sometimes serialized depending on the criticality of its application and interchangeable with other, standard components such as the belt of a conveyor, motor of a pump unit, or a bearing.

**Component ID**  A unique identification (ID) number used for tracking purposes and often physically attached to the component. The component ID is the lowest level in the asset hierarchy. Synonymous with *component number*.

**Component Number**  A component designation, usually structured by system, group, or serial number. Synonymous with *component ID*.

**Compound**  A distinct substance, formed by the combination of two or more elements in definite proportions by weight, possessing physical and chemical properties different from those of the combining elements.

**Compressibility**  The change in volume of a unit volume of a fluid when subjected to a unit change of pressure.

**Compression Packing**  Packing that accomplishes sealing by being deformed pressure.

**Compressor**  A device or machine that increases the pressure of a gas or vapor by increasing the gas density and delivering the fluid against the connected system resistance.

**Computer**  Any device capable of processing information to produce a desired result. No matter how large or small they are, computers typically perform their work in three well-defined steps:

1. Accepting input.
2. Processing the input according to predefined rules (programs).
3. Producing output.

There are several ways to categorize computers:

- Class – ranging from microcomputers to supercomputers
- Generation – first through fifth generation
- Mode of processing – analog versus digital

| | |
|---|---|
| **Computer Aided Design (CAD)** | Software used by architects, engineers, drafters and artists to create precision drawings or technical illustrations. CAD software can be used to create two-dimensional (2-D) drawings or three dimensional (3-D) models. |
| **Computer Aided Engineering (CAE)** | A broad term, used by the electronic design automation industry, for the use of computers to design, analyze and manufacture products and processes. CAE includes Computer Aided Design (CAD) and Computer Aided Manufacturing (CAM), which is the use of computers for managing manufacturing processes. |
| **Computer Aided Manufacturing (CAM)** | The use of computer technology to generate data to control part, or all, of a manufacturing process. |
| **Computer Aided Process Planning (CAPP)** | Software-based systems that aid manufacturing engineers in creating a process plan to manufacture a product whose geometric, electronic, and other characteristics have been captured in a CAD database. CAPP systems address such manufacturing criteria as target costs, target lead times, anticipated production volumes, availability of equipment, production routings, opportunity for material substitution, and test requirements. |
| **Computer Aided Software Engineering (CASE)** | The use of object-oriented programming, and other techniques, to streamline generation of programming code. |
| **Computer Integrated Manufacturing (CIM)** | A variety of approaches in which computer systems communicate or interoperate over a local-area network. Typically, CIM systems link management functions with engineering, manufacturing, and support operations. In a plant, CIM systems may control the sequencing of production operations, operation of automated equipment and conveyor systems, transmit manufacturing instructions, capture data at various stages of the manufacturing or assembly process, facilitate tracking and analysis of test results and operating parameters, or any combination of these. |
| **Computer Numerical Control (CNC)** | A technique that allows the control of motion, in an accurate and programmable manner, through the use of a dedicated computer within a numerical control unit that has the capability for local data input such that machine tools are freed from the need for "hard-wired" controllers. |

| | |
|---|---|
| **Computer Program** | A series of instructions or statements, in a form acceptable to a computer, that are designed to cause the computer to execute an operation or operations. |
| **Computerized Maintenance Management System (CMMS)** | A software package used to assist in the asset and work management functions. It tracks work orders, equipment histories, and preventive/predictive maintenance schedules. Usually it is integrated with support systems such as inventory control, purchasing, accounting, and manufacturing, and controls maintenance and warehouse activities. |
| **Computerized Process Simulation** | Use of computer simulation to facilitate sequencing of production operations, analysis of production flows, and layout of manufacturing facilities. |
| **Concurrent Engineering (CE)** | A cross-functional, team-based approach in which the product and the manufacturing process are designed and configured within the same time frame, rather than sequentially. Ease and cost of manufacturability, as well as customer needs, quality issues, and product-life-cycle costs are taken into account earlier in the development cycle. Fully configured concurrent engineering teams include representation from marketing, design engineering, manufacturing engineering, and purchasing, as well as supplier and even customer companies. |
| **Condition Assessment** | A process for judging the condition of operating equipment through a detailed, collective evaluation of condition measurements including anomalies from known condition, trends, and departures from previous characteristics. |
| **Condition Based Maintenance (CBM)** | Direction of maintenance actions based on indications of asset health as determined from non-invasive measurement of operation and condition indicators. CBM allows preventative and corrective actions to be optimized by avoiding traditional calendar or run based maintenance. |
| **Condition Based Maintenance Cost** | The costs associated with condition based maintenance activities. |
| **Condition Monitoring (CM)** | The continuous or periodic measurement and interpretation of data to indicate the condition of an item in order to determine the need for maintenance. |

| | |
|---|---|
| **Conditional Probability of Failure** | The probability that a failure will occur in a specific period provided that the item concerned has survived to the beginning of that period. |
| **Conduction** | Heat transfer from molecule to molecule, or atom to atom, not requiring movement of the substance. |
| **Conductivity – Electrical** | The ability of a material to conduct electric current. It is expressed in terms of the current per unit of applied voltage. It is the reciprocal of Resistivity. |
| **Conductivity – Thermal** | A measure of the ability of a material to conduct heat. |
| **Conductor – Electrical** | A material which carries electrical current (movable electric charges). Copper and aluminum are good, commonly used conductors. |
| **Conductor Complex Impedance** | The sum of a conductor's resistance, captive impedance, and inductive impedance. Accurate measurement of this "complex" conductor impedance allows degradation in a motor to be detected and addressed prior to motor failure. |
| **Cone – Bearing** | The inner ring of a tapered roller bearing. |
| **Confidence Interval** | An estimate of the interval between two statistics that includes the true value of the parameter with some probability. This probability is called the confidence level of estimate. The typical levels used are 90%, 95% and 99%. |
| **Confidence Level (CL)** | Statistical likelihood (probability) that a random variable lies within the confidence interval of an estimate. |
| **Configuration** | The arrangement and contour of the physical and functional characteristics of a system, equipment and related hardware or software. |
| **Configuration Audit** | A review of the product against the engineering specifications to determine whether the engineering documentation is accurate, up-to-date, and representative of the components, subsystems, or systems being produced. |

**Configuration Baseline** — The configuration documentation formally approved and released at a specific time during a system's or configuration item's life cycle. Configuration baselines, plus approved changes from those baselines, constitute the current configuration documentation.

**Configuration Change** — A change in the approved configuration of an item after formal establishment of its related baseline. The effects of a configuration change may involve configuration item modification, correction of documentation, and other item descriptions.

**Configuration Control** — The systematic proposal, justification, evaluation, coordination, approval (or disapproval) of proposed changes, and the implementation of all approved changes in the configuration of an item or component.

**Configuration Control Board (CCB)** — A group normally composed of technical and business decision makers to approve or reject proposed changes to the form, fit, or function of an item. Synonymous with *configuration review board*.

**Configuration Documentation** — Documentation identifying an asset's form, fit or function, such as drawings, operations and maintenance procedures, technical manuals, and specifications.

**Configuration Item** — An aggregation of system elements (hardware, firmware, or software and any of their discrete portions) that is designated for separate configuration management.

**Configuration Management (CMII)** — A discipline applying technical and administrative direction and surveillance in identifying and documenting the functional and physical characteristics of a configuration item, controlling changes to those characteristics, and recording and reporting changes to processing and implementation status.

**Configuration Manager** — The role of applying technical and administrative direction and surveillance in identifying and documenting the functional and physical characteristics of a configuration item, controlling changes to those characteristics, and recording and reporting changes.

**Confined Space** — An area with limited access and a potential respiratory hazard requiring a special permit to enter.

| | |
|---|---|
| **Conflict Resolution** | A process of resolving disagreements in a manner acceptable to all parties. |
| **Conformance** | An affirmative indication or judgment that a product or service has met the requirements of a relevant specification, contract or regulation. |
| **Connection – Electrical** | That part of a circuit that has negligible impedance and joins components and devices together. |
| **Consensus** | A state in which all the members of a group support an action or decision, even if some of them don't fully agree with it. |
| **Consequence** | Something that follows from an action or condition. The relation of a result to its cause. |
| **Consignment Stock** | Inventoried items that are physically in the storeroom and managed by storeroom personnel but which are still owned by the original supplier. |
| **Constant Bandwidth Filter** | A band-pass filter whose bandwidth is independent of center frequency. |
| **Constant Failure Rate** | A failure rate wherein the probability of failure does not change over time. |
| **Constant Horsepower Motor** | A term used to describe a multispeed motor in which the rated horsepower is the same for all operating speeds. |
| **Constant Torque Motor** | A multispeed motor for which the rated horsepower varies in direct ratio to the synchronous speeds. The output torque is essentially the same at all speeds. |
| **Constraint** | A bottleneck or limitation of the throughput of an asset or process. |
| **Consultant** | An individual, who has experience and expertise in applying tools and techniques to resolve process problems, who can advise and facilitate an organization's improvement efforts. |
| **Consumable** | Items, goods, or supplies that are used up after issuance from stores, or which are depleted or worn out by use, such as consumable paper products, safety wire, solder, tape, sanding disks, hand gloves, filters, etc. |

| | |
|---|---|
| **Consumer** | The external customer to whom a product or service is ultimately delivered. Synonymous with *end user*. |
| **Contact Ultrasonic** | An ultrasonic method that places the transducer directly in contact with the outside of the targeted component. The signal is passed into the material to a receiver that measures the characteristic of the echo which indicates discontinuities within the object. |
| **Contaminant** | Any foreign or unwanted substance that can have a negative effect on system operation, life, or reliability. |
| **Contaminant Failure** | Any loss of performance due to the presence of contamination.<br><br>Two basic types of contamination failure are:<br>1. Perceptible – The gradual loss of efficiency or performance.<br>2. Catastrophic – Dramatic, unexpected failure. |
| **Contamination Control** | A broad subject that applies to all types of material systems. It is concerned with planning, organizing, managing, and implementing all activities required to determine, achieve, and maintain a specified contamination level. |
| **Continuous Flow Production** | Items are produced and moved from one processing step to the next one piece at a time. Each process makes only the one piece that the next process needs, and the transfer batch size is one. |
| **Continuous Improvement (CI)** | The ongoing improvement of products, services, or processes through incremental and breakthrough improvements. Synonymous with *continual improvement*. |
| **Continuous Improvement Hours** | The number of internal maintenance labor hours that are used to improve the current performance to an increased level. Improvements may be in safety, quality, environment, availability, output, or cost. |
| **Continuous Process Improvement (CPI)** | Provides a framework for organizations to make incremental process improvements, even in those processes that are considered to be in good operating condition. It is based on the philosophy that organizations can always make improvements. |

|  |  |
|---|---|
| **Continuous Quality Improvement (CQI)** | A philosophy and attitude for analyzing capabilities and processes and improving them repeatedly to achieve the objective of customer satisfaction. |
| **Continuous Replenishment Programs** | Arrangement with supplier companies in which the supplier monitors the customer's inventory and automatically replaces used materials, eliminating the need for purchase orders and related paperwork. |
| **Continuous Sampling Plan** | A sampling plan, intended for application to a continuous flow of individual units of product, which involves acceptance and rejection on a unit by unit basis and employs alternate periods of 100% inspection and sampling. The relative amount of 100% inspection depends on the quality of submitted product. Continuous sampling plans usually require that each time period of 100% inspection be continued until a specified number of consecutively inspected units are found to be clear of defects. |
| **Contract Employees** | Personnel who are hired through a third company (typically to perform a specific function or a narrowly defined activity). Benefits and management are usually the responsibility of the third party company. |
| **Contract Maintenance** | Maintenance work performed by contractors. |
| **Contractor** | An individual or company providing specific services under a contract for those services, tasks, or specific results. |
| **Contractor Maintenance Cost** | The total expenditures for contractors engaged in maintenance on site. Includes all contractor maintenance labor and materials costs for normal operating times as well as outages, shutdowns, or turnarounds. Includes contractors used for capital expenditures directly related to end-of-life machinery replacement, but does not include contractors used for capital expenditures for plant expansions or improvements. |
| **Contrast** | The difference in visibility, brightness, color or temperature between an indication and the surrounding surface. |

**Control**  State in which all special causes of variation have been removed from a process. Processes held in control are monitored, usually by means of a control chart, so that corrective action can be taken if special cause variation returns.

**Control Chart**  A chart, with upper and lower control limits, on which values of some statistical measure for a series of samples or subgroups are plotted. The chart frequently shows a central line to help detect a trend of plotted values toward either control limit. The control chart helps to control and reduce process variations.

**Control Circuit**  A circuit in a piece of equipment, or an electrical circuit, that carries the signal determining the control action, as distinct from the power used to energize the various components.

**Control Limits**  The boundaries of a process within specified confidence levels, expressed as the upper control limit (UCL) and the lower control limit (LCL). A line–control limit usually used for judging process stability.

**Control Plan**  A document that describes the required characteristics for the quality of a product or service, including measures and control methods. The plan identifies critical input or output variables and associated activities that must be performed to maintain control of the variation of process.

**Control Point**  The value of a controlled variable that an automatic controller operates to maintain.

**Control Process**  A process involving gathering process data, analyzing process data, and using this information to make adjustments to the process.

**Control System**  A system to guide or manipulate various elements in order to achieve a prescribed value or result.

**Control Valve**  A valve, usually of a globe or needle configuration, that is positioned by the output of a control system and powered by a pneumatic diaphragm, electric motor, or hydraulic cylinder.

**Controller**  A device or program that operates automatically to regulate a controlled variable.

**Convection**  The transfer of energy caused by the movement of fluids, either liquid or gas.

**Coolant**  A fluid used to remove heat.

**Coordination**  Daily adjustment of maintenance activities to achieve the best short-term use of resources or to accommodate changes in operational needs for service.

**Coprocessor**  A processor, distinct from the main microprocessor, which performs additional functions or assists the main microprocessor.

**Core – Filter**  The internal duct and filter media support.

**Core Competencies**  Set of skills or knowledge sets that enable an organization to provide the greatest level of value to its customers. They are any aspect of the business operation that delivers a strategic business advantage.

**Corona**  An electrical glow appearing around the surface of a charged conductor. When voltage on an electrical conductor, such as an antenna or transmission line, exceeds a threshold value, the air around it begins to ionize to form a blue or purple glow.

**Corona Discharge**  An electrical discharge characterized by a corona. Occurring when one of two conducting surfaces (such as electrodes) of differing voltages has a pointed shape, resulting in a highly concentrated electric field at its tip, that ionizes the air (or other gas) around it. Corona discharge can result in power loss in the transmission of electric power.

**Corporate Culture**  The set of important assumptions that members of an organization share. A system of shared values about what is important, and beliefs about how the organization works. These common assumptions influence the way the organization operates. Synonymous with *organizational culture*.

**Corporation**  An organization that is a legal entity with rights, privileges, and responsibilities that are separate from those of its owners, the shareholders.

| | |
|---|---|
| **Corrective Action** | A documented design, process, procedure, or material change implemented and validated to correct the cause of failure or design deficiency. |
| **Corrective Action Recommendation** | The full cycle corrective action tool that offers ease and simplicity for employee involvement in the corrective action/process improvement cycle. |
| **Corrective Action Review** | A review conducted to assess the progress of the corrective action plan implementation created during previous compliance reviews. |
| **Corrective Action Status Report** | A periodic report used to document the progress of corrective actions identified as a result of negative review findings or management reviews. |
| **Corrective Maintenance (CM)** | Maintenance tasks initiated as a result of the observed or measured condition of an asset or system, before or after functional failure, to correct the problem. Corrective maintenance can be planned or unplanned. Synonymous with *corrective work*. |
| **Corrective Maintenance Cost** | The cost associated with corrective maintenance activities. |
| **Correlation** | A measure of the relationship between two data sets of variables. |
| **Corrosion** | The gradual eating away of metallic surfaces as the result of oxidation or other chemical action. |
| **Corrosion Inhibitor** | Additive for protecting lubricated metal surfaces against chemical attack by water or other contaminants. There are several types of corrosion inhibitors. Polar compounds wet the metal surface preferentially, protecting it with a film of oil. Other compounds absorb water by incorporating it in a water-in-oil emulsion so that only the oil touches the metal surface. Another type of corrosion inhibitor combines chemically with the metal to present a non-reactive surface. |
| **Cost** | The expenditure of resources (usually expressed in monetary units) necessary to develop, acquire, or use a product or service over some defined period of time. |

**Cost – Benefit Analysis**  Quantitatively evaluating the costs and benefits of a particular decision, program, project, or activity. Considering categories of benefits and costs, measuring them, and totaling their effects over time. Cost-benefit analysis provides the ability to justify the cost in improving maintenance.

**Cost Analysis**  Review and evaluation of actual, or anticipated, cost data to operate and maintain an asset.

**Cost Avoidance**  Action taken to reduce future costs, such as replacing parts before they fail and cause damage to other parts. Cost avoidance may incur higher (or additional) costs in the short-run but the final, or life cycle, cost will be lower.

**Cost Center**  The smallest segment of an organization for which data on costs are collected and formally reported.

**Cost Control**  The application of procedures to monitor expenditures and performance against the progress of any activities (such as maintenance projects, or operations with projected completion) to measure variances from authorized budgets and allow effective action to be taken to achieve minimal cost overruns.

**Cost Driver**  The key elements of an activity that drives the majority of the cost.

**Cost Effectiveness**  A measure of the value received (effectiveness) for the resources expended (cost) on an activity or asset.

**Cost Metrics**  Metrics used to measures the effect of maintenance activities on cost. Examples include maintenance cost as a percent of replacement asset value; stores inventory value as a percent of replacement asset value; and contractor costs as a percent of total maintenance costs.

**Cost Of Goods Sold (COGS)**  The total costs of the process to manufacture the product (material, labor, utilities, and overhead) in a specified period.

**Cost Of Poor Quality (COPQ)**  The costs associated with providing poor quality products or services.

There are four categories of costs:

1. Internal Failure Costs – associated with defects found before the customer receives the product or service.
2. External Failure Costs – associated with defects found after the customer receives the product or service.
3. Appraisal Costs – incurred to determine the degree of conformance to quality requirements.
4. Prevention Costs – incurred to keep failure and appraisal costs to a minimum.

**Cost Of Quality (COQ)**  The cost incurred to provide product or service quality. Major sources of COQ are quality of design and conformance. Quality costs include preventive costs, appraisal costs, and failure costs (both internal and external).

**Cost Reduction**  A policy of cutting costs to improve profitability. May be implemented when an organization is having financial problems and must "tighten its belt." In some cases, the company is initiating a policy to eliminate waste and inefficiency.

**Cost Savings**  An action that will result in a smaller than projected level of costs to achieve a specific objective.

**Count Chart**  A control chart for evaluating the stability of a process in terms of the count of events of a given classification occurring in a sample.

**Coupling**  A straight connector for fluid lines.

**Crack**  A break, or rupture, usually V-shaped and relatively narrow and deep.

**Cracking**  The process whereby large molecules are broken down by the application of heat and pressure to form smaller molecules.

**Cracking Pressure**  The pressure at which a pressure-operated valve begins to pass fluid.

**Craft**  Personnel skilled in generalized or specialized areas, who are directly involved in performing maintenance work. Craft includes helpers, mechanics, maintenance technicians, electronic technicians, instrumentation/calibration technicians, and others.

**Craft Worker**  A maintenance person who is responsible for executing work assignments pertaining to maintenance activities (both hourly and salaried personnel). Examples of craft workers include:

- technicians
- trades people
- PdM technicians
- apprentices
- tool carriers
- wrench turners
- mechanics
- electricians
- helpers

Synonymous with *craftsperson* or *tradesperson*.

**Craft-Wage Head Count**  The number of maintenance personnel responsible for executing work assignments pertaining to maintenance activities. Includes the number of contractor personnel who are used to supplement routine maintenance. Headcount is measured in full-time equivalents (FTE).

**Crane**  A machine for lifting and/or moving a load with the hoisting mechanism an integral part of the machine. Cranes may be traveling, portable, or fixed type.

**Crazing**  Minute crack(s) on the surface of materials, such as ceramics and plastics, caused by different rates of expansion or contraction of different layers.

**Creep**  A gradual change in the dimension of a material under a mechanical load, high stress, and/or high temperatures.

**Criteria**  Standards, rules, or tests on which a judgment or decision can be based, or by which a product, service, result, or process can be evaluated.

**Critical Asset** — An asset that has been evaluated and classified as critical due to its potential impact on safety, environment, quality, production, and maintenance costs if failed.

**Critical Damping** — The smallest amount of damping required to return the system to its equilibrium position without oscillation.

**Critical Failure** — A loss of function that could have a direct adverse effect on operations or safety.

**Critical Frequency** — A particular resonant frequency at which damage or degradation in performance is likely.

**Critical Path** — In a project network diagram, the series of activities that determine the earliest completion of the project.

**Critical Path Method (CPM)** — An activity-oriented project management technique that uses arrow diagramming methods to arrange and sequence tasks with preceding and succeeding events to demonstrate the time (and cost) required to complete a project or task.

**Critical Processes** — Processes that present serious potential dangers to human life, health, and the environment or that risk the loss of very large sums of money or customers.

**Critical Spare** — Any item that is kept on-hand because it is considered essential to the overall reliability of the operation, has a high cost, a long lead time, or an impact on safety, environmental concerns, or operations.

**Critical Spares** — Spare parts that have high value, long lead times, or are of particularly high value to the critical assets on which they are used. They are carried to avoid excessive downtime in the event of a breakdown.

**Critical Speeds** — Any rotating speed which results in high vibration amplitudes. Often these are speeds which correspond to system natural frequencies.

**Critical System** — A system that is vital to continued operations, will significantly impact production, or has inherent risks to personnel safety or the environment if failed.

**Critical Thinking** — A rigorous and thorough approach to examining a situation in such a way that all relevant factors and information are considered, appropriately weighed as to their impact, and analyzed, leading to a logical conclusion.

**Criticality** — A measure of the importance of an asset relative to other assets. A term that refers to how often a failure will occur, how easy it is to diagnose, and whether it can be fixed. Priority rank of a failure mode based on some assessment criteria.

**Criticality Analysis** — A quantitative analysis of events and faults, and the ranking of these, in order, based on a weighted combination of the seriousness of their consequences and frequency of occurrence.

**Criticality Matrix** — A matrix which is used to assess the likelihood and consequences of failure of a system, functional location, or equipment, and allocates an overall level of risk in terms of potential impact on operations in terms of production, safety, environment, or cost.

**Critical-To-Quality (CTQ)** — The characteristics of a product or service that are essential to ensure customer satisfaction.

**Cross Functional Team (CFT)** — A team of employees representing different functional disciplines and/or different process segments who tackle a specific problem or perform a specific task, frequently on an ad hoc basis.

**Cross-Crafting** — The training and qualifying of a maintenance craftsperson in other craft jobs.

**Crossover Frequency** — In sinusoidal vibration testing, the unique forcing frequency at which the required displacement yields the desired acceleration, and vice versa.

**Cross-Talk** — Interference or noise in a sensor or channel that comes from another sensor or channel.

**Cross-Training** — Skill-development practices that require, or encourage, production workers and other employees to master multiple job skills, thus enhancing workforce flexibility.

**Cryogenics** — Processes that achieve and maintain very low temperatures through the use of special refrigeration and cooling techniques.

**Cultural Resistance** — A form of resistance based on opposition to the possible social and organizational consequences associated with change.

|  |  |
|---|---|
| **Culture** | A common set of values, beliefs, attitudes, perceptions and accepted behaviors shared by individuals within an organization. |
| **Culture Change** | A major shift in the attitudes, norms, sentiments, beliefs, values, operating principles and behavior of an organization. |
| **Current Asset Value (CAV)** | The current cost to replace an asset. Synonymous with *current cost*. |
| **Current Cost** | The current cost to replace an asset. Synonymous with *current asset value*. |
| **Current Finish Date** | The current estimate of the point in time when a scheduled activity will be completed that reflects any reported work progress. |
| **Current Good Manufacturing Practices (cGMPs)** | Regulations enforced by the U.S. Food and Drug Administration (FDA) for pharmaceutical producers to provide for systems that assure proper design, monitoring, and control of manufacturing processes and facilities. |
| **Current Limiter** | A device used to instantaneously limit the flow of excessive electrical current (fault current) in a circuit, thereby protecting expensive electrical equipment. Fault currents are typically caused by short circuits, lightning, or common power fluctuations. |
| **Current Start Date** | The current estimate of the point in time when a scheduled activity will begin reflecting any reported work progress. |
| **Current-Cost Method** | A method of inflation accounting which replaces historical cost accounting principles with current (replacement) cost as the basis for financial-statement measurements. |
| **Current-Frequency Response Test** | A test which measures current on the same winding of a motor at two different frequencies to produce an indicator (phase angle change) of internal defects. |
| **Custodian** | The person responsible for protecting and preserving the property assigned to him or her. |
| **Customer** | The person or organization that will use the project's product, service or result. |

| | |
|---|---|
| **Customer Delight** | The result of delivering a product or service that exceeds customer expectations. |
| **Customer Lead Time** | The time elapsed from receipt of an order until the finished product is shipped to the customer. |
| **Customer Relationship Management (CRM)** | A strategy used to learn more about customers' needs and behaviors in order to develop stronger relationships with them. It brings together information about customers, sales, marketing effectiveness, responsiveness, and market trends. It helps businesses use technology and human resources to gain insight into the behavior of customers and the value of those customers. |
| **Customer Satisfaction** | The result of delivering a product or service that meets customer requirements. |
| **Customer Service** | The provision of service to customers before, during, and after a purchase. Customer service is a series of activities designed to enhance the level of customer satisfaction – the customer's feeling that a product or service has met their expectation. |
| **Customer Supplier Partnership** | A long-term relationship between a buyer and supplier characterized by teamwork and mutual confidence. The supplier is considered an extension of the buyer's organization. The partnership is based on several commitments. The buyer provides long-term contracts and uses fewer suppliers. The supplier implements quality assurance processes so incoming inspection can be minimized. The supplier also helps the buyer reduce costs and improve product and process designs. |
| **Cut Set** | In fault tree analysis, any group of initiators which, if all occur, will cause the top event to occur. |
| **Cutting Fluid** | Any fluid applied to a cutting tool to assist in the cutting operation by cooling, lubricating, or other means. |
| **Cycle** | A single complete operation consisting of progressive phases, starting and ending at the neutral position. |

**Cycle Counting**  An inventory accuracy audit technique where inventory is counted on a cyclic schedule rather once a year. The main purpose of cycle counting is to identify items in error, thus triggering research, identification, and elimination of the cause of the errors.

**Cycle Time**  The time required to complete one cycle of an operation. It is elapsed time between the start and completion of a task or an entire process. For example, in order processing, it can be the time between receipt and delivery of an order.

**Cycle Time Reduction**  A method for reducing the amount of time it takes to execute a task, process, or build a product.

**Cyclic Damage**  An effect in which decreasing strength is a function of the total number of load occurrences and their magnitudes.

**Cylinder**  A device which converts fluid power into linear mechanical force and motion. It usually consists of a moveable element (such as a piston and piston rod, plunger rod, plunger, or ram) operating within a cylindrical bore.

# D

**D/A Converter** — A device that converts a digital signal (discrete values) into an analog voltage.

**Daily Schedule** — Jobs that are scheduled to be the worked the following day.

**Damper** — A device that deadens, restrains, or depresses.

**Damping – Mechanical** — The absorption of energy, as in viscous damping of mechanical energy. Examples include a dashpot or shock absorber.

**Damping – Vibration** — The dissipation of oscillatory or vibratory energy with motion or time.

**Data** — A set of collected facts. There are two basic types:
1. Measured or Variable Data
2. Counted or Attribute Data

**Data Acquisition System** — Any instrument or computer that acquires data from sensors via amplifiers, multiplexers, and any necessary analog to digital converters.

**Data Capture Rate** — The rate at which data or information packets are gathered by instruments and analyzers such as infrared cameras, vibration analyzers, and PLCs.

**Data Collection** — Gathering and recording of facts, changes, and forecasts for analysis.

**Data Collection Strategy** — A definitive plan for collecting quantitative and qualitative data, which includes type of data to be collected, the sources of the data, the method of collection, and the sampling frequency.

**Data Element** — A single piece of data that cannot be subdivided and still retain any meaning. Synonymous with *data item* and *data field*.

| | |
|---|---|
| **Data Entry** | The process in which an operator uses a keyboard or other device to input data directly into a system. |
| **Data File** | A collection of related data records, or application data values, organized in a specific manner and stored after, and separated from, the user program area. |
| **Data Logging** | Recording of data about events that occur in time sequence. |
| **Data Management** | The process by which the reliability, timeliness, and accessibility of an organization's database are assured. |
| **Data Mining** | An information extraction activity whose goal is to discover facts contained in databases. |
| **Data Warehouse** | A repository of an organization's electronically stored data. Data warehouses are designed to facilitate reporting and analysis. |
| **Database** | A file composed of records, each containing fields, together with a set of operations for searching, sorting, recombining and other functions. |
| **Date** | A term representing the day, month, and year of a calendar. In some instances, can include the time of day. |
| **DC Motor** | A motor using either generated or rectified direct current (DC) power. A DC motor is often used when variable-speed operation is required. |
| **Deadband** | In process control, a range in which an input signal may be varied without initiating a change in output signal. |
| **Deadman Control** | Mechanism designed to remove, or reduce, hazardous energy to a safe level in the absence of the kind of force exerted by a conscious operator. |
| **Deaerator** | A separator that removes air from the system fluid through the application of bubble dynamics. |
| **Debugging** | The process of finding and eliminating errors. |
| **Deburring** | Removing burrs left by machining, shearing, and casting operations. Deburring is usually done by grinding. |

**Decentralized Maintenance**

A maintenance organization in which multiple maintenance groups report to specific business or production organizations.

**Decentralized Stores**

Spare parts stores that are distributed throughout a facility, not centralized in one location.

**Decibel (dB)**

A measure of power gain or loss relative to an arbitrarily chosen power level. Equal to 10 times the logarithm of the ratio between output power and base-level power. As applied to speech and sound levels, dB is measurement of sound energy ratios.

**Decision**

A determination arrived at after review, investigation, interpretation, or consideration of facts and/or information.

**Decision Making**

Analyzing a problem to identify viable solutions and then making a choice among them.

**Decision Matrix**

A tool used to evaluate and prioritize a list of options. It can also be used to systematically identify, analyze, and rate the strength of relationships between sets of information. Synonymous with *Pugh matrix, decision grid, selection matrix or grid, problem matrix, problem selection matrix, opportunity analysis, solution matrix, criteria rating form, criteria-based matrix* and *ranking matrix*.

**Decision Support System (DSS)**

Computer software used to aid in decision making. May involve simulation programming, mathematical programming routines, and decision rules.

**Decision Tree**

A diagram that shows key interactions among decisions so they are understood by the decision maker. Branches of the tree represent either decisions or chance events. The diagram provides consideration of the probability of each outcome.

**Decomposition – Chemical**

The breaking up of compounds into smaller chemical forms through the application of heat, change in other physical conditions, or introduction of other chemical bodies.

| | |
|---|---|
| **Decomposition – Project** | Subdivision of the major project deliverables into smaller, more manageable components until the deliverables are defined in sufficient detail to support future project activities (planning, executing, controlling and closing). |
| **Decontamination** | Removal of a polluting or harmful substance from air, water, earth surface, etc. For example, the process of removing hazardous chemical contamination from objects or areas. |
| **Deductive Reasoning** | Forms of inference such that the conclusion must be true if the premises are true. Logicians contrast deduction with induction, in which the conclusion might be false even when the premises are true. Deduction has to do with necessity, induction with probability. |
| **De-Energized** | Free from any electrical connection to a source of potential difference and electric charge. Not having a potential different from that of the earth. Note: the term is used only with reference to current-carrying parts, which are sometimes energized (live). |
| **De Facto Authority** | Influence exercised regardless of formal authority, position or bureaucratic knowledge. |
| **Default Value – Computer** | The option taken by a computer in the event of the omission of a definite instruction or action. |
| **Default Value – Reset** | A value that is used when no value is provided in the instance document. Default values apply to attributes that are either empty or missing. |
| **Defect** | A condition that causes deviation from design or expected performance, leading to failure. A fault. Anything that erodes value, reduces production, compromises health, safety and environment or creates waste. |
| **Defective** | A unit of product that contains one or more defects with respect to the quality characteristic(s) under consideration. |
| **Defective Units Produced** | The number of unacceptable units produced during a time period (i.e., losses, rework, scrap, etc.). |
| **Defects Per Million Opportunities (DPMO)** | The actual number of defects occurring, divided by the total number of opportunities for a defect, multiplied by one million. Synonymous with *ppm (parts per million)*. |

| | |
|---|---|
| **Defects Per Unit (DPU)** | The total number of defects divided by the number of units or products. |
| **Deferred Maintenance** | Maintenance which can be postponed to some future date without further deterioration of equipment. |
| **Deficiency Tag** | An information tag hung by the requester of maintenance work to identify the equipment. |
| **Deflection** | The amount of the deviation from a straight line or plane when a force is applied to a press member. Generally used to specify allowable bending of bed, slide, or frame at rated capacity with load of predetermined distribution. |
| **Deformation** | Changes of shape under load. |
| **Degas** | Removing air from a liquid, usually by ultrasonic and/or vacuum methods. |
| **Degradation** | An irreversible process in one or more characteristics of an item involving either time, use, or an external cause. |
| **Degraded State** | State of an item whereby that item continues to perform a function to acceptable limits but lower than the specified values, or, continues to perform only some of its required functions. |
| **Degrees of Freedom – Mechanical** | The specific, defined modes in which a mechanical device or system can move. The number of degrees of freedom is equal to the total number of independent displacements or aspects of motion. A machine may operate in two or three dimensions but have more than three degrees of freedom. Freely suspended assemblies have six degrees of freedom, three mutually perpendicular and three rotational. |
| **Degrees of Freedom – Statistics** | The number of values in the final calculation of a statistic that are free to vary. Estimates of statistical parameters can be based upon different amounts of information or data. The number of independent pieces of information that go into the estimate of a parameter is called the degrees of freedom. |
| **Dehydrator** | A separator that removes water from the system fluid. |

|  |  |
|---|---|

**Delamination** — A failure mode for composite material. In laminated materials, repeated cyclic stresses or impacts can cause layers to separate with significant loss of mechanical toughness.

**Delamination Wear** — A complex wear process where a machine surface is peeled away, or otherwise removed, by forces of another surface acting on it in a sliding motion.

**Delay** — Performing an action later than originally scheduled. For example, postponing maintenance tasks to a later time.

**Delay Time** — Any delay or wait time. This could be any task, activity, project, craft or crane wait time.

**Delay Time – Signal** — The average time delay in a linear system having a nonlinear phase characteristic. The concept is based upon the determination of the time delay which minimizes the mean-square error between input and output signals.

**Delegation** — Assigning a project, or task, and setting the corresponding expectations. Delegation can occur formally or informally. The former is usually associated with performance related issues and personnel development. The latter occurs under impromptu circumstances, that is, when the opportunity presents itself.

**Delighter** — A feature of a product or service that a customer does not expect but that gives pleasure when received.

**Delimited Files** — Files in which the different data fields are set apart by a delimiting character, usually a comma.

**Delimiter** — A special character that sets off, or separates, individual items in a program or set of data. Special characters often used include commas, semi-colons, tabs, and paragraph marks.

**Deliverable** — Any unique and verifiable product, result, or capability to perform a service, which must be produced to complete a process, phase, or project.

**Delphi Method** — A form of participative expert judgment. An iterative, anonymous, interactive technique using survey methods to derive consensus on work estimates, approaches, and issues.

**Demerit Chart**  A control chart for evaluating a process in terms of a demerit, or quality score. A weighted sum of counts of various classified nonconformities.

**Deming Cycle**  Concept of a continuous improvement cycle consisting of plan-do-check-act (PDCA), which is used to show the need for interaction among all departments to improve quality. Synonymous with *Deming wheel*, *Plan-Do-Check-Act*, and *Shewhart cycle*.

**Deming Prize**  Award, given annually, to organizations that, according to the award guidelines, have successfully applied companywide quality control based on statistical quality control and are committed to maintaining it in the future. Although the award is named in honor of W. Edwards Deming, its criteria are not specifically related to Deming's teachings.

There are three separate divisions for the award:

1. The Deming Application Prize.
2. The Deming Prize for Individuals.
3. The Deming Prize for Overseas Companies

The award process is overseen by the Deming Prize Committee of the Union of Japanese Scientists and Engineers in Tokyo.

**Deming's Fourteen Points**  W. Edwards Deming's fourteen management points to increase quality and productivity when practiced.

They are:

1. Create constancy of purpose for improving product or services.
2. Adopt new philosophy.
3. Cease dependence on inspection to increase quality level.
4. End the practice of awarding business on price alone. Work with a single supplier as a partner.
5. Improve constantly and forever every process for planning, production, and service.
6. Institute training on the job.
7. Adopt and institute leadership.
8. Drive out fear.
9. Break down barriers between staff areas.
10. Eliminate slogans, exhortations, and targets for the workforce.
11. Eliminate numerical quota for the workforce and numerical goals for management.
12. Remove barriers that rob people of pride of workmanship.
13. Institute a vigorous program of education and self improvement.
14. Put everybody in the organization to work to accomplish the transformation.

**Democratic Management Style**  Participative management approach in which the manager and employees make decisions jointly.

**Demographic Analysis**  A set of methods that allow measurement of the dimensions and dynamics of populations.

**Demonstrated**  That which has been measured by the use of objective evidence gathered under specified conditions.

**Demulsibility**  The ability of a fluid, that is insoluble in water, to separate from water with which it may be mixed in the form of an emulsion.

**Density**  The ratio of the mass of a substance to its volume.

| | |
|---|---|
| **Dependability** | The collective term used to describe availability performance and its influencing factors (Reliability, Maintainability, Performance, Maintenance, Support, Performance). |
| **Dependent Failure** | Failure which is caused by the failure of an associated item(s). |
| **Dependent Tasks** | Tasks that are related such that the beginning or end of one task is contingent on the beginning or end of another. |
| **Deployment** | To arrange in a position of readiness, or to move strategically or appropriately, such as to deploy a new process. |
| **Deposits** | Oil-insoluble materials that result from oxidation and decomposition of lube oil and contamination from external sources and engine blow-by. These can settle out on machine or engine parts. Examples are sludge, varnish, lacquer and carbon. |
| **Depot Maintenance** | High-level maintenance performed on equipment requiring major overhaul or substantial or complete rebuilding. This term is usually used in the U.S. Department of Defense. |
| **Depreciation** | The gradual conversion of the cost of a tangible capital asset or fixed asset (excluding land which has unlimited life) into an operational expense (called depreciation expense) over the asset's estimated useful life. |
| **Depth Filter** | A filter medium that retains contaminants primarily within tortuous passages. |
| **Derating** | The use of a part for conditions less severe than its rated conditions so as to obtain improved reliability. |
| **Design** | Creation of the description of a product or service, in the form of drawings, diagrams, or other methods, to provide detailed information on how to build the product or perform the service. |
| **Design Adequacy** | The probability that a system will successfully accomplish its design objectives given that it is operating within design specifications. |

**Design Development Phase**  The phase of a designer's basic services, which include developing structural, mechanical, and electrical drawings, specifying materials, and estimating the probable cost of construction.

**Design Engineering**  The systematic and creative application of scientific and mathematical principles to practical ends such as the design, manufacture, and operation of efficient and economical structures, machines, processes, and systems.

**Design Failure Modes and Effects Analysis (DFMEA)**  An analysis process using failure modes and effects analysis (FMEA) to identify and evaluate the relative risk associated with a particular design.

**Design For Assembly (DFA)**  The practice in which ease and cost of assembly are emphasized as an integral part of the design process.

**Design for Logistics**  The practice in which physical handling and distribution of a manufactured product are emphasized as an integral part of the design process.

**Design For Manufacturability (DFM)**  The practice in which ease and cost of manufacturing, as well as quality-assurance issues, are emphasized as an integral part of the design process.

**Design for Procurement**  A practice in which product designers work effectively with suppliers and sourcing personnel to identify and incorporate technologies, or designs, which can be used in multiple products, facilitating the use of standardized components to achieve economies of scale and assure continuity of supply.

**Design for Quality**  The practice in which quality assurance and customer perception of product quality are emphasized as an integral part of the design process.

**Design for Recycling and Disposal**  The practice in which ultimate disposal and recycling of the manufactured product are emphasized as an integral part of the design process.

**Design For Reliability (DFR)**  Improving the reliability of an asset, process or product by using reliability analysis techniques (e.g., DFMEA, FMEA, and RBD) to design out potential problems during the development phase.

| | |
|---|---|
| **Design For Six Sigma (DFSS)** | A systematic methodology utilizing tools, training, and measurements to design products and processes that meet customer expectations and can be produced at Six Sigma quality levels. |
| **Design Life** | The period for which the design is intended to perform its requirements. After the target period, the item is expected to be discarded because it ceases to function or becomes too expensive to repair. |
| **Design Limit** | The operational limit of an asset or product beyond which it not required to function properly. |
| **Design of Experiments (DoE)** | An experimental design methodology that enables process designers to determine optimum product/process parameters by conducting a limited number of experiments involving combinations of variables. The usual objective is to determine which variables in a complex process are most critical for quality control or which can be most easily changed to reduce overall process variance. |
| **Design Record** | Engineering requirements typically contained in various formats. Examples include engineering drawings, math data, and referenced specifications. |
| **Design Review** | A formal, documented, comprehensive and systematic examination of a design to evaluate its capability to meet specified requirements, identify problems, and propose solutions. |
| **Design Specifications** | Precise measurements, tolerances, materials, in-process and finished-product tests, quality control measures, inspection requirements, and other specific information that precisely describes how the work is to be done. |
| **Desired Performance** | The level of performance desired by the owner, or user, of a physical asset or system. |
| **Desired Quality** | The additional features and benefits a customer discovers when using a product or service that leads to increased customer satisfaction. If it is missing, a customer may become dissatisfied. |
| **Desorption** | The opposite of absorption or adsorption. In filtration, it relates to the downstream release of particles previously retained by the filter. |

| | |
|---|---|
| **Detergent** | In lubrication, either an additive or a compounded lubricant having the property of keeping insoluble matter in suspension, preventing its deposition where it would be harmful. A detergent may also redisperse deposits already formed. |
| **Deterioration Rate** | The rate at which an item approaches a departure from its functional standard. |
| **Deterministic Model** | A mathematical model that, given a set of input data, produces a single output or a single set of outputs. |
| **Deterministic Vibration** | A vibration whose instantaneous value at any future time can be predicted by an exact mathematical expression. Sinusoidal vibration is the classic example. Complex vibration is less simple (two or more sinusoids). |
| **Development Phase** | The project's life cycle phase during which project planning and design typically occur. Synonymous with *planning phase*. |
| **Deviation – Control** | The difference, at any instant, between the value of the controlled variable and the desired set point or control value. |
| **Deviation – Mathematical** | In numerical data sets, the difference or distance of an individual observation or data value from the center point (often the mean) of the set distribution. |
| **Device** | Any electrical or mechanical instrument, or combination, used in an asset or system to produce or provide data. |
| **Device Driver** | A software component that permits a computer system to communicate with a device. In most cases, the driver also manipulates the hardware in order to transmit the data to the device. |
| **Dew Point** | The temperature at which vapor starts to condense. |
| **Diagnosis** | The activity of discovering the cause(s) of deficiencies. The process of investigating symptoms, collecting and analyzing data, and conducting experiments to test theories to determine the root cause(s) of deficiencies. |
| **Dial Indicator** | A measuring device, equipped with a readout dial, used to align shafts. |

**Diaphragm Pump** — A pump that uses a positive displacement design rather than centrifugal force to move water through the casing, delivering a specific amount of flow per stroke, revolution, or cycle.

**Diaphragm Valve** — A control valve that responds to the signal from a controller and uses air pressure as the activating force.

**Dielectric** — An insulating substance such as oil, liquid nitrogen, or paper.

**Dielectric Strength** — A measure of the ability of an insulating material to withstand electric stress (voltage) without failure. Fluids with high dielectric strength (usually expressed in volts or kilovolts) are good electrical insulators.

**Differential Pressure – Indicator** — An indicator which signals the difference in pressure between any two points of a system or component.

**Differential Pressure – Pump** — The difference between the outlet pressure and the inlet pressure.

**Digital Control System (DCS)** — A system that uses digital signals and a digital computer to control a process.

**Digital Signal Processing (DSP)** — Digital signal processing (DSP) is the technology of extracting information, in the time domain or in the frequency domain, from signals that are digitized at a specified sampling rate. The primary goal of DSP is to extract information about the operation of machinery or a device with applications to monitoring, diagnostics, control, predictive maintenance, estimation of performance related parameters, pattern recognition, and sensor validation.

**Dilution** — Reducing the concentration of a chemical mixture.

**Direct Cost** — Cost identified with a specific, final cost objective. Not necessarily limited to items that are incorporated into the end product or service as labor or material.

**Direct Current (DC)** — A current that flows in only one direction in an electric circuit.

| | |
|---|---|
| **Direct Labor** | Labor identified with a specific, final cost objective. For example, manufacturing direct labor includes fabrication, assembly, inspection, and test for constructing the end product. |
| **Direct Purchase Item** | An item that is not an inventoried item, typically a one-time purchase. |
| **Directional Control Servo-Valve** | A directional control valve which modulates flow or pressure as a function of its input signal. |
| **Directional Control Valve** | A valve whose primary function is to direct or prevent flow through selected passages. |
| **Directive** | Written communication that initiates or prescribes action, conduct, or procedure. |
| **Dirt Capacity** | The weight of a specified artificial contaminant which must be added to the influent to produce a given differential pressure across a filter at specified conditions. Used as an indication of relative service life. Synonymous with *dust capacity* and *contaminant capacity*. |
| **Dirty Power** | Fluctuations in the voltage or current on a power system which can affect the performance and operation of electrical equipment. |
| **Disassemble** | To open an asset or system and remove a number of parts or subassemblies in order to repair or replace them. |
| **Discard Task** | The removal and disposal of items or parts. |
| **Discipline** | An area of technical expertise or specialty. |
| **Discount Rate** | The interest rate used in calculating the present value of future cash flow. |
| **Discounted Cash Flow (DCF)** | A method of investment analysis in which future cash flows are converted, or discounted, to their value at the present time. The rate of return for an investment is that interest rate at which the present value of all related cash flow equals zero. |
| **Discounting** | The process of reducing the amounts of a stream of future cash flows to their present value. |

| | |
|---|---|
| **Discrete Data** | Data that are counted instead of measured. Only a finite number of values are possible. |
| **Discrete Fourier Transform** | A procedure for calculating discrete frequency components (filters) from sampled time data. |
| **Discrete Manufacturing** | A manufacturing process in which the output is individually countable, or is identifiable by serial numbers, and is measurable in distinct units rather than by weights or volume. |
| **Discrete Variable** | A variable whose possible values are from a finite set. |
| **Discrete Work Package** | A short-term task with a definite start and end point that can be used to measure work performance or earned value. |
| **Discretionary Fixed Cost** | Fixed costs that can be eliminated at management's discretion in a relatively short period of time (for example, some administrative salaries, research and development, new systems development). |
| **Discriminant Analysis** | A statistical technique that is used to classify observations into two or more groups if the samples have a known grouping. |
| **Disk – Computer** | A round, flat piece of flexible plastic coated with a magnetic material that can be electrically influenced to hold information recorded in digital (binary) form. |
| **Dispersant** | In lubrication, a term used interchangeably with detergent. An additive, usually nonmetallic (ashless), which keeps fine particles of insoluble materials in a homogeneous solution. Hence, particles are not permitted to settle out and accumulate. |
| **Dispersion** | The degree of data scatter, usually about an average value such as the median or mean. |
| **Displacement** | A change of position or distance, usually measured from mean position or position of rest. Usually applies to uniaxial, less often to angular, motion. |
| **Displacement Transducer** | A transducer whose output is proportional to the distance between it and the measured object, usually a shaft. |

|  |  |
|---|---|
| **Display** | The visual output device of a computer. Either a CRT or LCD based video display, or a gas plasma–based flat-panel display. |
| **Disposable** | An item, such as a filter element, intended to be discarded and replaced after one service cycle. |
| **Disruptive Management Style** | A management approach in which the manager tends to destroy the unity of an organization or team, be an agitator, and cause disorder. |
| **Disruptive Stress** | The physical or mental stress a person feels that threatens, frightens, angers or worries them, resulting in poor or ineffective performance. |
| **Dissatisfiers** | The features or functions a customer expects that either are not present or are present but not adequate. Also pertains to employees' expectations. |
| **Dissipation** | The concept of a dynamical system where important mechanical modes, such as waves or oscillations, lose energy over time, typically due to the action of friction or turbulence. The lost energy is converted into heat, raising the temperature of the system. |
| **Dissolved Gases** | Those gases that enter into solution with a fluid, and are neither free nor entrained gases. |
| **Dissolved Inorganic** | A water contaminant that includes calcium and magnesium dissolved from rock formations, gases such as carbon dioxide that ionize in water, silicates leached from sandy river beds or glass containers, ferric and ferrous ions from rusty iron pipes, chloride and fluoride ions from water treatment plants, phosphates from detergents, and nitrates from fertilizers. |
| **Dissolved Organics** | A water contaminant that may include pesticides, herbicides, gasoline, and decayed plant and animal tissues. Also includes the plasticizers leached out of plumbing lines, styrene monomers from fiberglass-reinforced storage tanks, deionized resin materials, and carbon from activated carbon filters. |

**Distance Learning** — Method of providing education and training to an individual or group of individuals in which the learner(s) is not in the same room as the provider (teacher, professor, instructor) of the information. Requires the use of one or more of a variety of techniques and media to present the subject matter and can include television, satellite broadcasts, correspondence, e-mail, computer-based training, CD ROM, videotape, Web-based delivery, or any combination thereof.

**Distillation** — A process for separating liquid mixtures based on differences in their volatilities.

**Distillation Method – Tribology** — A method involving distilling the fluid sample in the presence of a solvent that is miscible in the sample but immiscible in water. The water distilled from the fluid is condensed and segregated in a specially-designed receiving tube or tray, graduated to directly indicate the volume of water distilled.

**Distortion** — An alteration of the original shape (or other characteristic) of an object, image, sound, waveform or other form of information or representation. Distortion is usually unwanted.

**Distributed Control System (DCS)** — A type of automated control system that is distributed throughout a machine to provide instructions to different parts of the machine. Instead of having a centrally located device controllin all machines, each section of a machine has its own computer that controls the operation. A DCS is commonly used in manufacturing equipment and utilizes input and output protocols to control the machine.

**Distribution – Statistical** — The amount of potential variation in the outputs of a process, typically expressed by its shape, average or standard deviation.

**Distribution Management System** — A system utilized to determine optimal quantities of each product to be made at each plant and to be distributed to each warehouse, such that manufacturing and distribution costs are minimized and customer demands are met.

**Dividend** — A payment made to shareholders at a stated amount per share, usually quarterly, and usually in cash. Can also be a benefit result of a program or project.

**Division of Labor**  Specifically assigning persons to various activities by categories of labor, skill, or expertise.

**DMADV Process**  An acronym for Define, Measure, Analyze, Design and Verify. DMADV is a data driven quality strategy for designing products and processes, and is an integral part of a Six Sigma quality initiative.

**DMAIC Process**  An acronym for Define, Measure, Analyze, Improve and Control. It is a process for continued improvement. It is systematic, scientific, and fact based. This closed-loop process eliminates unproductive steps, often focuses on new measurements, and applies technology for improvement.

**Document**  A medium, and the information recorded thereon, that generally has permanence and can be read by a person or a machine.

**Document Control**  System to control and disseminate documentation in an orderly fashion.

**Document Management System (DMS)**  A server-based network facility designed for the storage and handling of an organization's documents. A DMS is built around a central library known as a repository and typically supports controlled access, version tracking, cataloging, search capabilities, and the ability to check documents in and out electronically.

**Documentation**  Collection of reports, information, records, references, and other data for distribution and archival purposes.

**Documented Training**  Training for which there is documented proof that the person received the training. Examples include employee orientation, safety, craft skills, and continuing education.

**Dodge-Romig Sampling Plans**  Plans for acceptance sampling developed by Harold F. Dodge and Harry G. Romig. Four sets of tables were published in 1940:
1. Single Sampling Lot Tolerance
2. Double Sampling Lot Tolerance
3. Single Sampling Average Outgoing Quality Limit
4. Double Sampling Average Outgoing Quality Limit Tables

**Down** — Out of service, usually due to breakdown, unsatisfactory condition, or production scheduling.

**Downtime** — The amount of time an asset is not capable of running. It is the sum of scheduled and unscheduled downtime.

**Downtime Event** — An event when the asset is down and not capable of performing its intended function.

**Drift** — The slow variation of a performance characteristic, such as gain, frequency, or power output. Usually, drift is only significant when measuring low-level signals (a few millivolts) over long periods of time or in difficult environmental conditions.

**Driving Forces** — Forces that tend to change a situation in desirable ways.

**Dropping Point** — The temperature at which grease passes from a semisolid to a liquid state. The change of state is the result of the breakdown of thickening agents.

**Drum** — A container usually with a capacity of 55 U.S. gallons.

**Dual Probe** — A transducer set consisting of displacement and velocity transducers. It combines measurement of shaft motion relative to the displacement transducer with velocity of the displacement transducer to the absolute motion of the shaft.

**Duane Plot** — A predictive model measuring product reliability over time as yielded by the manufacturing process. Predominantly used on electronic equipment to measure the effectiveness of a reliability program.

**Duct** — A passageway to conduit, made of sheet metal or other suitable material, used for conveying air or gas at low pressures.

**Ductility** — The property that permits permanent deformation before fracture by stress in tension.

**Due Date** — The required date for the delivery of an item, or completion of a task, project, or activity.

| | |
|---|---|
| **Dunnage** | Any material, such as boards, planks, blocks, plastics, or metal bracing, that is used in transportation and storage to support and secure supplies, protect them from damage, or for convenience in handling. |
| **Duplex Filter** | An assembly of two filters with valving for selection of either or both filters. |
| **Duplex Pump** | A reciprocating, positive displacement pump having two pumping cylinders. In direct-acting steam driven pumps, having two power cylinders, each connected to one of two pumping cylinders. |
| **Durability** | The ability of an item to perform a required function under given conditions of use and maintenance, until a limiting state is reached. |
| **Duration – Vibration** | In a shock pulse: its length. In a classical pulse: the length between instants when the amplitude is greater than 10% of the peak value. |
| **Duration – Work** | The number of work periods required to complete an activity, usually expressed as hours, workdays, or workweeks. |
| **Dust Tight** | Constructed so that dust will not enter an enclosing case under specified test conditions. |
| **Dustproof** | Constructed or protected so that dust will not interfere with successful operation. |
| **Duty Cycle** | The fraction of time during which a device or system will be active, or at full power. |
| **Dynamic Joint** | A joint intended to accommodate expansion and contraction movements of the structure. Synonymous with *expansion joint*. |
| **Dynamic Motion** | Movement, as compared with nonmoving or static position. Dynamic motion is sensed with displacement or velocity pickups or with accelerometers. |
| **Dynamic Range** | The ratio of a specified maximum level of a parameter, such as power, current, voltage, or frequency, to the minimum detectable value of that parameter. |

**Dynamic Signal Analyzer**  Vibration analyzer using digital signal processing and Fast Fourier Transform (FFT) to display vibration frequency components. May also display the time domain and the phase spectrum. Usually interfaced to a computer.

**Dynamometer**  A device which places a load on the motor to accurately measure its output torque and speed by providing a calibrated dynamic load. Helpful in testing motors for nameplate information, and an effective device in measuring efficiency.

# E

**Early Finish Date**  The earliest possible point in time when the uncompleted portions of an activity (or project) can end based on network logic and any schedule constraints. May change as the project progresses or as changes are made to the project plan. Used in the critical path method.

**Early Start Date**  The earliest possible point in time when the uncompleted portions of an activity (or project) can begin based on network logic and any schedule constraints. May change as the project progresses or as changes are made to the project plan. Used in the critical path method.

**Earnings**  The difference between the revenues and the expenses for an accounting period. Synonymous with *net income*.

**Eccentricity – Mechanical**  Variation of shaft surface radius when referenced to the shaft's true geometric centerline. Out-of-roundness

**E-Commerce**  Conducting business transactions between businesses, or between businesses and consumers, usually over the Web, or through automated means.

**Economic Life**  The total length of time that an asset is expected to remain actively in service before it is expected that it would be cheaper to replace the equipment rather than continuing to maintain it.

In practice, equipment is more often replaced for other reasons, including:

- because it no longer meets operational requirements for efficiency, product quality, comfort, etc.
- because newer equipment can provide the same quality and quantity of output more efficiently

**Economic Order Quantity (EOQ)**  A fixed order quantity that minimizes the total of carrying (holding) and order preparation costs under conditions of certainty and independent demand.

**Economy of Scale**  The decline of the total cost per unit as volume increases, often due to the existence of fixed costs.

**Eddy Current** — Electrical current generated (and dissipated) in a conductive material (often a rotor shaft) when it intercepts the electromagnetic field of a displacement or proximity probe.

**Eddy Current Testing (ECT)** — A method of detecting and isolating surface or near surface flaws. Eddy current testing (ECT) is used on electrically conductive materials. It is based on inducing small, circular currents in a test specimen. ECT uses high frequency AC current in a primary coil to generate a magnetic field, which in turn induces circular eddy currents in the material of interest. The ECT device obtains a signal by disruption of the eddy current due to a flaw in the test equipment, causing a change in the opposing magnetic field, and thus changing the coil impedance. ECT data consist of impedance plane signatures, inductive reactance, and resistance.

**Effect** — What results after an action has been taken. The expected or predicted impact when an action is to be taken or is proposed.

**Effectiveness** — Degrees to which objectives are achieved and the extent to which targeted problems are resolved. In contrast to efficiency, effectiveness is determined without reference to costs. Whereas efficiency means doing the thing right, effectiveness means doing the right thing.

**Effects** — The consequences of failures. Loss of one or more of the component's functions.

**Efficiency** — The ability to produce a desired effect or product with a minimum of effort, expense, or waste. It is the quality or fact of being efficient. Doing the thing right.

**Efficiency – Energy** — The ratio between useful work performed and the energy expended in producing it. It is the ratio of output power divided by input power.

**Efficient** — A term describing a process that operates effectively while consuming the minimum amount of resources (such as labor and time).

| | |
|---|---|
| **Effluent** | A liquid waste generated from a manufacturing or treatment process. The waste may be untreated, partially treated, or completely treated before discharge into the environment. Also, the fluid leaving a component. |
| **Effusivity – Thermal** | The ability of heat to escape from a body, expressed as a characteristic of that body. It is the square root of the product of thermal conductivity, mass density and specific heat. |
| **Elastomer** | A rubber-like material that, when compressed and then released, will return to 90% of its original shape in less than five seconds. |
| **Electric Circuit** | The path followed by electrons from a power source (generator or battery) through an external line (including devices that use the electricity) and returning back to the source through another line. |
| **Electrical Ground** | A conducting connection between an electrical circuit or equipment and the earth. A connection to establish ground potential. |
| **Electrical Hazard** | A dangerous condition such that contact or equipment failure can result in electric shock, arc flash burn, thermal burn, or blast. |
| **Electrical Motor** | A device that converts electrical energy into useful mechanical power. |
| **Electrical Relay** | A device designed to produce sudden predetermined changes, in one or more electrical circuits, after the appearance of certain conditions in the controlling circuit. |
| **Electrical Signal Analysis (ESA)** | The procedure of capturing an electro-mechanical system's current and voltage signals and analyzing them to detect various faults. |
| **Electricity** | The flow of electrons through a conducting medium. |
| **Electromagnetic Radiation** | Vibrating electrical and magnetic fields in the form of waves which travel at the speed of light. |
| **Electromagnetic Spectrum** | The range of electromagnetic radiation of varying wavelengths from gamma rays to radio waves. |

| | |
|---|---|
| **Electromechanical Relay** | An electrical relay in which the designed response is excited by a relative mechanical movement of elements under the action of a current in the input circuit. |
| **ElectroMotive Force (EMF)** | Potential causing electricity to flow in a closed circuit. |
| **Electronic Data Interchange (EDI)** | A standard for automated exchange of business documents. Using EDI, purchasers and suppliers can exchange digital paperwork including purchase orders, invoices, and other business documents, as well as performing electronic funds transfers. |
| **Electronic Funds Transfer** | The electronic transfer of payments. |
| **Electrostatic Separator** | A separator that removes contaminant from dielectric fluids by applying an electrical charge to the contaminant that is then attracted to a collection device of different electrical charge. |
| **Element** | A general term used to refer to a subset of a system, subsystem, assembly, or subassembly. A part or group of parts comprising a system. |
| **Element – Cartridge** | The porous device that performs the actual process of filtration. |
| **E-mail** | The exchange of text messages and computer files over a communications network, such as a local area network or the Internet, usually between computers or terminals. |
| **Emergency** | A condition requiring immediate, corrective action for safety, environmental, or economic risk, caused by equipment breakdown. |
| **Emergency Change** | Changes that must be accomplished without delay to avoid serious compromise of asset effectiveness, or to correct a hazardous condition which may result in fatal or serious injury to personnel or extensive damage or destruction of equipment. |
| **Emergency Maintenance Task** | A maintenance task carried out in order to avert an immediate safety or environmental hazard, or to correct a failure. Also emergency repairs or emergency work. |

**Emergency Spares**          Replacement parts required for critical assets/systems that are kept in reserve in anticipation of outages caused by man-made or natural disaster.

**EMI Noise**                 Disturbances to electrical signal caused by electromagnetic interference (EMI).

**Emission Spectrometer**     A device that works on the basis that atoms of metallic and other particular elements emit light at characteristic wavelengths when they are excited in a flame, arc, or spark.

**Emissivity**                A property of a material that describes its ability to radiate energy in comparison to a blackbody at the same temperature. Emissivity values range from zero to one (blackbody = 1).

**Emittance**                 The property of a material in a situation that describes its ability to radiate energy in comparison to a blackbody at the same temperature.

**Empirical**                 A formula derived or guided from experience, experiment, or observation alone.

**Employee**                  General term for an employed wage earner or salaried worker.

**Employee Involvement**      A practice within an organization whereby employees regularly participate in making decisions on how their work areas operate, including making suggestions for improvement, planning, goal setting, and monitoring performance.

**Empowered Natural Work Teams**   Teams that share a common workspace and/or responsibility for a particular process or process segment. Typically, such teams have clearly defined goals and objectives related to day-to-day production activities, such as quality assurance and meeting production schedules, as well as authority to plan and implement process improvements. Unlike self-directed teams, empowered work teams typically do not assume traditional "supervisory" roles.

**Empowerment** — A condition whereby employees have the authority to make decisions and take action in their work areas without prior approval. For example, an operator can stop a production process if he or she detects a problem, or a customer service representative can send out a replacement product if a customer calls with a problem.

**Emulsibility** — The ability of a non-water-soluble fluid to form an emulsion with water.

**Emulsion** — An intimate mixture of oil and water, generally of a milky or cloudy appearance. Emulsions may be of two types: oil-in water (where water is the continuous phase) and water-in-oil (where water is the discontinuous phase).

**Enclosed Space** — A working space, such as a manhole, vault, tunnel, or shaft, that has a limited means of egress or entry, that is designed for periodic employee entry under normal operating conditions, and that under normal conditions does not contain a hazardous atmosphere, but that may contain a hazardous atmosphere under abnormal conditions.

**Enclosure** — A surrounding case that protects equipment from its environment, and protects personnel against contact with the enclosed equipment.

**End Cap** — A ported or closed cover for the end of a filter element.

**End User** — Person or group for whom a product or service is developed.

**Endorsement** — Written approval that signifies personal understanding and acceptance of the thing approved and recommends further endorsement by higher levels of authority if necessary. Signifies authorization if endorsement of commitment is by a person with appropriate authority.

**Endoscope** — Device for viewing the interior of objects.

**End-to-End Testing** — A test in which the sensor, control unit, and executing element of a control loop are all examined.

**Energized**  Electrically connected to a source of potential difference, or electrically charged so as to have a potential significantly different from that of earth in the vicinity. Synonymous with *live*.

**Energy**  That which does work or is capable of doing work.

**Energy Storage**  Reserving electric energy for later use to avoid blackouts or fluctuations in power. Storage methods include batteries, flywheels, pumped hydropower and superconducting magnetic energy storage.

**Engineering Analysis**  The process of applying engineering concepts to the design of a product or process, including tests such as heat transfer analysis, stress analysis, or analysis of the dynamic behavior of the system being designed.

**Engineering Change (EC)**  A revision to a blueprint or design released by engineering to modify or correct a part, component, equipment, or system.

**Engineering Standard**  Design or test guidelines intended to promote the design, production, and test of part, component, or product in a manner that promotes standardization, ease of maintenance, consistency, adequacy of test procedures, versatility of design, ease of production and field service, and minimization of the number of different and special tools required.

**Engineering Support**  Engineers, and engineering technicians, who support the maintenance function. This could include facilities and equipment maintenance, the support of small capital projects, overhauls, and high-expense projects that are maintenance oriented. It includes process improvement engineers, reliability maintenance engineers, etc.

**Engineering Units**  Units that are decided upon by an individual user, or by agreement among users. Examples include inches/second, mm/s, g, Hz, etc.

**Engineering Work Order**  The prime document used to initiate an engineering investigation, engineering design activity or engineering modifications to an item of equipment. A control document from engineering, authorizing changes or modifications to a previous design or configuration.

| | |
|---|---|
| **Enterprise** | A venture characterized by innovation, creativity, dynamism, and risk. An enterprise can consist of one project, or may refer to an entire organization. It usually requires several of the following attributes:<br>• Flexibility<br>• Initiative<br>• Problem solving ability<br>• Independence<br>• Imagination |
| **Enterprise Integration (EI)** | A broad implementation of information technology to link various functional units within a business enterprise. On a wider scale, it may also integrate strategic partners in an enterprise configuration. In a manufacturing enterprise, enterprise integration (EI) may be regarded as an extension of computer integrated manufacturing (CIM) that integrates financial or executive decision-support systems with manufacturing, tracking and inventory systems, product-data management, and other information systems. |
| **Enterprise Integration Team** | A team that coordinates, makes, and implements enterprise-level decisions. |
| **Enterprise Process** | A process defined for application across an organization which applies to all people, assets, and organizations and is both documented and under enterprise configuration control. |
| **Enterprise Process Management System** | A system used for the life cycle management of enterprise processes. Used to manage resources and assets to develop, control, and improve Enterprise Processes. |
| **Enterprise Process Owner** | The person responsible for maintaining the Enterprise Process, and who serves as a champion of the assigned enterprise process throughout its lifecycle. |
| **Enterprise Process Performance Review** | A verification of process performance against established goals. |

| | |
|---|---|
| **Enterprise Resource Planning (ERP)** | A software system comprised of a single, or integrated, suite of applications to manage a variety of functional areas, including materials management, supply chain management, production, sales and marketing, distribution, finance, field service and human resources. It also provides information linkages to help companies monitor and control activities in geographically dispersed operations. |
| **Environment** | The aggregate of all external and internal conditions (such as humidity, temperature, radiation, magnetic and electric fields, shock vibration, etc.) natural, manmade, or self-induced, that influence the form, performance, reliability, or survival of an asset or facility. |
| **Environmental Concerns** | Plant activities that impact air, water, and soil quality as driven by local, state, and federal laws (e.g., EPA) and corporate conscience. Actions required to minimize, or prevent and monitor, adverse effects on the environment in order to comply with the laws. Corrective actions taken to clean up, handle, and remove contaminated water or soil. |
| **Environmental Consequences** | A failure mode, or multiple failures, that have environmental consequences if in breach of any corporate, municipal, regional, national or international environmental standard or regulation which applies to the physical asset or system under consideration, |
| **Environmental Contaminant** | Pollutants, or contaminants, that can affect health, which are found in air, water, food, and soil. |
| **Environmental Stress Screening** | A post-production process in which 100% of produced units are subjected to stresses more severe than anticipated in service. The object is to precipitate latent defects into recognizable failures, so that the failed unit does not proceed further in production or reach the customer. |
| **Environmental Testing** | Subjecting a sample of products to a simulation of anticipated storage, transport, and service environments (such as vibration, shock, temperature, altitude, humidity, etc.). |
| **Equipment** | An item having a complete function apart from being a substructure of a system. |

| | |
|---|---|
| **Equipment Life** | The useful or design life of equipment. |
| **Equipment Lifetime** | A span of time over which the equipment is expected to fulfill its intended purpose. |
| **Equipment Maintenance Strategy** | The choice of routine maintenance tasks and the timing of those tasks, designed to ensure that an item or equipment continues to fulfill its intended functions. |
| **Equipment Operating Procedures** | Formal written procedures for starting up, running, and shutting down equipment. |
| **Equipment Repair History** | A chronological list of failures, repairs, and modifications on assets. Synonymous with *maintenance history* and *maintenance record*. |
| **Equipment Use** | A measure of the accumulated hours, cycles, distance, throughput, etc., that an asset has performed its function. |
| **Equivalent Distillation Capacity** | A normalized value for comparing refinery processing units based on processing intensity. Determined from criteria developed by Solomon Associates. |
| **Ergonomics** | The science that matches human capabilities, limitations, and needs with that of the work environment. |
| **Erosion** | Loss of material, or degradation of surface quality, through friction or abrasion from moving fluids, cavitation or by particle impingements. |
| **Error** | The difference between the indicated, and the true value, of a variable being measured. |
| **Error Proofing** | A process used to prevent errors from occurring, or to immediately point out a defect as it occurs. |
| **Escalation** | An increase in price, especially due to inflation. |
| **Estimated Replacement Value (ERV)** | The estimated cost to replace plant's current production equipment and support facilities at current market value. Synonymous with *replacement asset value (RAV)*. |
| **Estimate-To-Complete (ETC)** | The funding needed to complete a project. |

| | |
|---|---|
| **Ethernet** | The standard for local communications networks developed jointly by Digital Equipment Corp., Xerox, and Intel. Ethernet baseband coaxial cable transmits data at speeds up to ten megabits per second. Ethernet is used as the underlying transport vehicle by several upper-level protocols, including TCP/IP. |
| **Ethics** | The practice of applying a code of conduct, based on moral principles, to day-to-day actions to balance what is fair to individuals or organizations against what is right for society. |
| **European Foundation for Quality Management (EFQM) Excellence Model** | A non-prescriptive framework based on nine criteria and 32 sub-criteria used to assess an organization's progress towards excellence. Based on the premise that customer satisfaction, people satisfaction, and impact on society are achieved through excellence that focuses on policy and strategy, people management, resources, and processes. |
| **European Quality Award** | An award established and presented by the European Foundation for Quality Management (EFQM) to those European firms that demonstrate commitment to business excellence in accordance with the EFQM Excellence Model. |
| **Eutectic** | An alloy used to form the melting point of a fuse. It is frequently silver or tin based. |
| **Evaporation** | The conversion of a liquid into vapor, usually by means of heat. |
| **Evaporation Loss** | The loss of fluid volume or weight as a result of the evaporation of the fluid. |
| **Event Tree Analysis (ETA)** | A process designed to determine the probability of an event based on the outcomes of each event in the chronological sequence leading up to it. |
| **Evident Failure** | A failure mode that, on its own, becomes apparent to the users of the asset under normal operating circumstances. |
| **Examination** | A comprehensive inspection with measurement and physical testing to determine the condition of an item. |

| | |
|---|---|
| **Excavation** | Man-made cavity or depression in the earth's surface, including its sides, walls, or faces, formed by earth removal and producing unsupported earth conditions. |
| **Excess Inventory** | Any inventory in the system that exceeds the minimum amount necessary to achieve the optimal stocking level. |
| **Excitation** | An external force (or other input) applied to a system that causes the system to respond in some way. |
| **Expansion Joint** | A joint or coupling designed so as to permit an endwise movement of its parts to compensate for expansion or contraction. |
| **Expectations** | Customer perceptions about how an organization's products and services will meet their specific needs and requirements. |
| **Expected Quality** | The minimum benefit a customer expects to receive from a product or service. |
| **Expeditor** | The role of facilitating the delivery of materials, tools, or services, to the job site in advance of task scheduling. |
| **Expense** | Items chargeable as overhead or operating cost. Funds used in the maintenance, repair, or restoration of property to an operating state without extending the useful accounting life or materially increasing its value. Expenditures associated with relocations and rearrangements of property. |
| **Experiment** | A test under controlled conditions that is made to demonstrate a known truth, examine the validity of a hypothesis, or determine the efficacy of something previously untried. |
| **Experimental Design** | A formal plan that details the specifics for conducting an experiment, such as which responses, factors, levels, blocks, treatments and tools are to be used. |
| **Experimental Training Techniques** | Training that is hands-on and provides the recipients of the training with the opportunity to experience, in some manner, the concepts that are being taught. |

| | |
|---|---|
| **Expert System** | A system that uses knowledge and is rule-based since it is easy to codify expertise using rules. The area of Expert Systems investigates methods and techniques for constructing man-machine systems with specialized problem-solving expertise. Expertise consists of knowledge about a particular domain, understanding of problems in this domain, and skill at solving some of these problems. |
| **Explosion Proof** | An item designed and constructed to withstand an internal explosion without creating an external explosion or fire. |
| **Exponential Distribution** | A distribution that is a class of continuous probability distributions. They describe the times between events in a Poisson process, i.e. a process in which events occur continuously and independently at a constant average rate. Probably the most widely known and used distribution in reliability evaluation of systems. |
| **Express Warranty** | A promise spoken or written in an agreement. |
| **External Audit** | An audit performed by a person from outside the organization. |
| **External Customer** | A person or organization that receives a product, service, or information but is not part of the organization supplying it. |
| **External Document** | A document provided by an external source such as the government, customer, manufacturer, industry association, or statutory and regulatory authority that contains information that is necessary to instruct, control, or regulate work activities. Synonymous with *document of external origin*. |
| **External Failure** | Nonconformance identified by external customers. |
| **External Validation** | Using benchmarking as a way to ensure that an organization's current practices are comparable to those being used by outside organizations. |
| **Extranet** | An exclusionary Internet-like network that securely connects customers and suppliers to a corporate or plant intranet in order to access information deemed sharable by the intranet operators. |

**Extreme Pressure Additive**   A lubricant additive that prevents sliding metal surfaces from seizing under conditions of extreme pressure. At the high local temperatures associated with metal-to-metal contact, an extreme pressure additive combines chemically with the metal to form a surface film that prevents the welding of opposing asperities, and the consequent scoring that is destructive to sliding surfaces under high loads. Reactive compounds of sulfur, chlorine, or phosphorus are used to form these inorganic films.

**Extreme Pressure Lubricants**   Lubricants that impart to rubbing surfaces the ability to carry appreciably greater loads, without excessive wear or damage, than would be possible with ordinary lubricants.

# F

**F Distribution** — A right-skewed distribution used most commonly in analysis of variance (ANOVA).

**F – Test** — Any statistical test in which the test statistic has an F-distribution if the null hypothesis is true. It is most often used when comparing statistical models that have been fit to a data set in order to identify the model that best fits the population from which the data were sampled.

**Facilitation** — The design and management of processes and structures that enable groups to succeed in their missions.

**Facilitator** — An individual responsible for creating a favorable environment that will enable a team or group to reach consensus, or achieve its goal, by bringing together the necessary tools, information, and resources to get the job done.

**Facilities** — The buildings, equipment, machinery, systems, and grounds that are used for production or service.

**Facility Management** — The activity of looking after an organization's equipment, buildings, infrastructure, and processes.

**Facility Management Plans** — Annual plans summarizing the asset maintenance and investment work expected to be completed within a fiscal year.

**Fail-Safe** — A device or feature which, in the event of failure, responds in a way that will cause no harm, or minimal harm, to other devices and personnel.

**Fail-Safe Design** — A design property of an item that prevents its failures from causing harm to the system or personnel.

**Failure** — The inability of an asset to perform its required function.

**Failure Analysis** — The process of analyzing a failure to determine its cause and to put measures in place to prevent future problems.

**Failure Cause**     Initiator of the process by which deterioration begins, resulting ultimately in failure.

**Failure Code**     A code typically entered against a work order in a CMMS which indicates the cause of a failure from a predefined list (e.g., lack of lubrication, metal fatigue, etc.).

**Failure Consequences**     The way (s) in which the effects of a failure mode or a multiple failure matter (e.g., evidence of failure, impact on safety, the environment, operational capability, direct and indirect repair costs).

**Failure Cost**     The cost associated with an asset or service failure resulting from not conforming to requirements or user needs.

**Failure Data Analysis**     A type of reliability analysis used to study a variety of fields, practices, and disciplines, including mechanical, chemical, electrical, electronic, materials, and human failures. It is useful for maintenance planning, developing cost-effective replacement strategies, spare parts forecasting, and warranty analysis.

**Failure Distribution Parameters**     Each failure distribution requires the definition of distribution parameters in order to properly analyze reliability data. These parameters vary with the specific failure distributions in use, and can include the mean, lower and upper bound, standard deviation, characteristic life, and shape factor, among others.

**Failure Effect**     The consequence of a failure mode on the function or status of an item.

**Failure Finding Interval**     The frequency with which a failure finding task is performed.

**Failure Finding Task**     A routine maintenance task, normally an inspection or testing task, designed to determine whether an item or component has failed. A failure finding task should not be confused with an on-condition task which is intended to determine whether an item is about to fail. Failure finding tasks are sometimes referred to as functional tests.

**Failure Management Policy**     A generic term that encompasses on-condition tasks, scheduled restoration, scheduled discard, failure-finding, run-to-failure, and one-time changes.

**Failure Mechanism**  The physical, chemical, electrical, thermal, or other process which results in a failure.

**Failure Mode**  The manner in which a failure occurs.

**Failure Mode Analysis (FMA)**  A procedure to determine which malfunction symptoms appear immediately before or after failure of a critical parameter in a system. After all the possible causes are listed for each symptom, the product or process is re-designed to eliminate the problems.

**Failure Mode Effects Analysis (FMEA)**  A procedure in which each potential failure mode in every sub-item of an item is analyzed to determine its effect on other sub-items and on the required function of the item.

**Failure Mode Effects and Criticality Analysis (FMECA)**  A procedure that is performed after a failure mode effects analysis to classify each potential failure effect according to its severity and probability of occurrence.

**Failure Pattern**  A graph showing conditional probability of failure against operating age for an item or group of items. Synonymous with *pattern of failure*.

**Failure Probability**  The probability that failure occurs before time t. If R (t) is the reliability and F (t) is the failure probability at time t, then F(t) = 1 - R(t).

**Failure Rate**  The number of failures of an asset over a period of time (per unit measurement of life). Failure rate is considered constant over the useful life of an asset. It's normally expressed as the number of failures per unit time and is the inverse of mean time between failure (MTBF). The failure rate is denoted by Lambda ($\lambda$).

**Failure Rate Curve**  A graphic representation of the relationship of the life of an asset or product versus the probable failure rate. Synonymous with *bathtub curve*.

| | |
|---|---|
| **Failure Reporting, Analysis, and Corrective Action System (FRACAS)** | A formal management information system that supports five distinct and basic functions:<br>1. Recording data about failures.<br>2. Reporting data to engineers/analysts for data analysis.<br>3. Analysis to discover failure causes.<br>4. Documenting corrective action plans.<br>5. Checking on corrective action adequacy if any further action is required. |
| **Failure-Finding Task** | A scheduled task that seeks to determine if a hidden failure has occurred or is about to occur. |
| **Fan** | A device that produces a pressure difference in air to move it. |
| **Fast Fourier Transform (FFT)** | A procedure for calculating discrete frequency components from sampled time data. It is a special case of the discrete Fourier Transform, where the number of samples is constrained to a power of 2 for speed. |
| **Fatality Rate** | The number of fatalities in an organization during the year. For US companies it is reported on section G of OSHA form 300A. |
| **Fatigue – Human** | A decline in performance due to physiological or psychological tiring. |
| **Fatigue – Metal** | The progressive and localized structural damage that occurs when a material is subjected to cyclic loading. The maximum stress values are less than the ultimate tensile stress limit, and may be below the yield stress limit of the material. |
| **Fatigue Chunks** | Three-dimensional particles exceeding 50 microns indicating severe wear of gear teeth. |
| **Fatigue Platelets** | Normal particles between 20 and 40 microns found in gear box and rolling element bearing oil samples observed by analytical ferrography. A sudden increase in the size and quantity of these particles indicates excessive wear. |
| **Fault** | To function in an undesired manner, or to be unable to function in a desired manner, regardless of cause. |

**Fault Isolation** — The process of determining the location of a fault to the extent necessary to affect repair.

**Fault Isolation Time** — The amount of time that is necessary to gain access to and isolate the fault (or failures).

**Fault Localization** — The process of determining the approximate location of the fault.

**Fault Tree Analysis (FTA)** — A logical, structured process that can help identify potential causes of system failures (faults) before the failures actually occur. A deductive analysis begins with a general conclusion then attempts to determine the specific causes of this conclusion. This is often described as a "top down" approach.

**Feedback – Customer** — Communication from customers about how delivered products or services compare with customer expectations.

**Feedback Signal** — A signal responsive to the value of the controlled variable in an automatic controller. The signal is returned as an input of the control system and compared with the reference signal, obtaining an actuating signal that returns the controlled variable to the desired value.

**Ferrography** — An analytical method of assessing machine health by quantifying and examining ferrous wear particles suspended in the lubricant or hydraulic fluid.

**Fiber Optics** — A technology for the transmission of light beams along optical fibers. A light beam, such as that produced in a laser, can be modulated to carry information.

**Fiber-Reinforced Concrete** — Concrete-based construction material used in applications such as slabs and overlays, precast products, structural beams and girders, and shotcrete applications. Glass, polyethylene, or steel fibers are added to other concrete ingredients. Additions of the fibers results in an improvement in strength, shock resistance, and ductility.

**Fidelity** — The accuracy with which an electronic system reproduces the sound or image of its input signal.

**Field** — A location in a record in which a particular type of data is stored.

**Fieldbus** — The name of a family of industrial computer network protocols used for real-time distributed control, now standardized as IEC 61158.

**File** — A complete, named collection of information such as a program, a set of data used by a program, or a user created document. A file is the basic unit of storage that enables a computer to distinguish one set of information from another.

**Film Strength** — The property of a lubricant that acts to prevent scuffing or scoring of metal parts.

**Filter – Electrical** — An electronic device to pass certain frequencies (pass band) but block other frequencies (stop band). Classified as low-pass (high-stop), high-pass (low-stop), band-pass or band-stop.

**Filter – Mechanical** — Any device or porous substance used as a strainer for cleaning fluids by removing suspended matter.

**Filter Efficiency** — A method of expressing a filter's ability to trap and retain contaminants of a given size.

**Filter Element** — The porous device which performs the actual process of filtration.

**Filter Head** — An end closure for the filter case or bowl that contains one or more ports.

**Filter Housing** — A ported enclosure that directs the flow of fluid through the filter element.

**Filter Life Test** — A type of filter capacity test in which a clogging contaminant is added to the influent of a filter, under specified test conditions, to produce a given rise in pressure drop across the filter or until a specified reduction of flow is reached. Filter life may be expressed as test time required to reach terminal conditions at a specified contaminant addition rate.

**Filter Media, Depth** — Porous materials which primarily retain contaminants within a tortuous path, performing the actual process of filtration. Synonymous with *depth filter media*.

| | |
|---|---|
| **Filter Media, Surface** | Porous materials which primarily retain contaminants on the influent face, performing the actual process of filtration. Synonymous with *surface filter media*. |
| **Filtration** | The physical or mechanical process of separating insoluble particulate matter from a fluid, such as air or liquid, by passing the fluid through a filter medium that will not allow the particulates to pass. |
| **Filtration Ratio** | The ratio of the number of particles greater than a given size in the influent fluid to the number of particles greater than the same size in the effluent fluid. Synonymous with *Beta ratio*. |
| **Financial Budget** | A summary of a major aspect of an organization's economic status which is used as a fiscal control technique. |
| **Finished-Goods Turn Rate** | A measure of asset management that is typically calculated by dividing the value of total annual shipments at plant cost (for the most recent full year) by the average finished-goods inventory value. Plant cost includes material, labor, and plant overhead. |
| **Finite Element Analysis (FEA)** | A mathematical method for analyzing stress. Finite element analysis is used in product design software to conduct graphical on-screen analysis of a model's reactions under various load conditions. |
| **Fire Point** | The lowest temperature at which a liquid will ignite and achieve sustained burning when exposed to a test flame in accordance with ASTM D 92, Standard Test Method for Flash and Fire Points by Cleveland Open Cup Tester. |
| **Fire Protection Rating** | The time, in minutes or hours, that materials and assemblies used as opening protection have withstood a fire exposure as established in accordance with test procedures of NFPA 252. |
| **Fire Resistance** | Refers to the properties of materials or design to resist the effects of fire to which the material or structure may be expected to be subjected. |
| **Fire Wall** | A security system intended to protect an organization's network against external threats, such as hackers, coming from another network, such as the Internet. |

**Fire-Resistant Fluid** — A lubricant used especially in high-temperature or hazardous hydraulic applications, such as steel mills and underground mining.

Three common types of fire-resistant fluids are:

1. Water-Petroleum Oil Emulsions – the water prevents burning of the petroleum constituent
2. Water-Glycol Fluids – a glycol antifreeze in water
3. Non-Aqueous Fluids – low volatility fluids such as phosphate esters, silicones, and halogenated hydrocarbon-type fluids

**Firmware** — Software routines stored in read-only memory (ROM). Unlike random access memory (RAM), read-only memory stays intact even in the absence of electrical power. Startup routines and low-level input/output instructions are stored in firmware. It falls between software and hardware in terms of ease of modification.

**First Aid Incident** — Injuries or illnesses that do not meet the minimum threshold to be recordable.

**First Law of Thermodynamics** — Energy in a closed system is constant. It cannot be created or destroyed.

**First Order Vibration** — Rotating machine vibration caused by shaft unbalance. Frequency in hertz (Hz) is calculated by shaft RPM/60. Synonymous with *1x vibration*. Additional orders, 2x, 3x... 36x, etc. are caused by other mechanisms.

**First-In-First-Out (FIFO)** — An inventory valuation method in which costs of materials are transferred in chronological order of receipt.

**First-Pass Yield (FPY)** — The percentage of finished products that meet all quality-related specifications at a final test point. When calculating yield for components, the percentage that meets all quality-related specifications at a critical test point. FPY is calculated by dividing the units entering the process, minus the defective units, by the total number of units entering the process. Synonymous with *quality rate*.

| | |
|---|---|
| **Fishbone Analysis** | A graphical technique that can be used to identify and arrange the causes of an event, problem, or outcome. The hierarchically relationships between the causes according to their level of importance or detail are illustrated on branches or "bones". Synonymous with *cause-and-effect* and *Ishikawa analysis*. |
| **Fishbone Diagram** | The diagram resulting from a fishbone analysis. Synonymous with *cause-and-effect* and *Ishikawa diagram*. |
| **Fit – Mechanical** | Range of mechanical interferences, tightness, or clearance looseness, resulting from mating parts which have given sizes, allowances, and tolerances. |
| **Five S** | A structured program to reduce waste and optimize productivity utilizing five terms beginning with "S" to create a workplace suited for visual control and lean operations. |

The Five S' are:

1. Sort (Seiri) – to separate needed tools, parts, and instructions from unneeded materials and to remove the latter.
2. Set-in-Order (Seiton) – to neatly arrange and identify parts and tools for ease of use.
3. Shine (Seisō) – to conduct a cleanup campaign.
4. Standardize (Seiketsu) – to conduct Seiri, Seiton, and Seiso at daily intervals to maintain a workplace in perfect condition.
5. Sustain (Shitsuke) – to form the habit of always following the first four S'.

Many organizations have added a sixth S... *Safety*.

| | |
|---|---|
| **Five Whys** | A technique for discovering the root causes of a problem and showing the relationship of causes by repeatedly asking, *Why?* |
| **Fixed Bearing** | A bearing which positions a shaft against axial movement in either direction. |
| **Fixed Costs** | Costs that are independent of short-term variations in output of the system under consideration. |
| **Fixed Displacement Pump** | A pump in which the displacement per cycle cannot be varied. |

**Flammable**  A combustible that is capable of easily being ignited and rapidly consumed by fire. Flammables may be solids, liquids or gases exhibiting these qualities.

**Flammable Liquid**  Any liquid having a flash point below 100°F (37.8°C) and having a vapor pressure not exceeding an absolute pressure of 40 psi (276 kPa) at 100°F (37.8°C).

**Flash Point**  The minimum temperature of a liquid at which sufficient vapor is given off to form an ignitable mixture with the air near the surface of the liquid, or within the vessel used, as determined by the appropriate test procedure and apparatus specified in NFPA 30, Flammable and Combustible Liquids Code, 1.7.4.

**Flexible Machining Centers**  Automated machining equipment that can be rapidly reprogrammed to accommodate small-lot production of a variety of product or component configurations.

**Flexible Manufacturing System (FMS)**  Automated manufacturing equipment and/or cross-trained work teams that can accommodate small-lot production of a variety of product or part configurations. From an equipment standpoint, a flexible manufacturing system is typically a group of computer-based machine tools with integrated material handling that is able to produce a family of similar parts.

**Flexible Rotor**  A rotor which operates close enough to, or beyond, its first bending critical speed for dynamic effects to influence rotor deformations. Rotors which cannot be classified as rigid rotors are considered to be flexible rotors.

**Flexible Workforce**  A workforce whose members are cross-trained and whose work rules permit assignment of individual workers to different tasks.

**Float Switch**  A device used to start and stop a pump based on preset levels.

**Floating Bearing**  A bearing designed or mounted so as to permit axial displacement between shaft and housing.

**Floating-Point Number**  A number represented by a mantissa and an exponent according to a given base. The mantissa is usually a value between 0 and 1. Ordinary scientific notation uses floating-point numbers with 10 as the base.

| | |
|---|---|
| **Flow Control Valve** | A valve whose primary function is to control flow rate. |
| **Flow Fatigue Rating** | The ability of a filter element to resist a structural failure of the filter medium due to flexing caused by cyclic differential pressure. |
| **Flow Rate** | The volume, mass, or weight of a fluid passing through any conductor per unit of time. |
| **Flowchart** | A graphical representation of the steps in a process. Flowcharts are drawn to better understand processes. |
| **Flowcharting** | The depiction, in a diagram format, of the inputs, process actions, and outputs of one or more processes within a system. |
| **Flowmeter** | A device which indicates either flow rate, total flow, or a combination of both. |
| **Fluid** | A general classification including liquids and gases. |
| **Fluid Compatibility** | Suitability of filtration medium and seal materials for service with the fluid involved. |
| **Fluid Friction** | Friction due to the viscosity of fluids. |
| **Fluid Opacity** | The ability of a fluid to allow light to pass through it. |
| **Fluid Power** | The energy transmitted and controlled through use of a pressurized fluid. |
| **Flushing** | A fluid circulation process designed to remove contamination from the wetted surfaces of a fluid system. |
| **Focal Plane Array (FPA)** | An infrared imaging system that uses a matrix type detector such as 240x320 pixels. It can be either radiometric or qualitative. |
| **Focus Group** | A group, usually eight to ten persons, that is invited to discuss an existing or planned product, service or process. |
| **Foolproofing** | A method of making a product or process immune to errors on the part of the user or operator. Synonymous with *error proofing* and *mistake proofing*. |

**Footprint** — The amount of floor or table space taken up by a unit or object.

**Force Feed Lubrication** — A system of lubrication in which the lubricant is supplied to the bearing surface under pressure.

**Force Field Analysis** — A technique for analyzing the forces that aid or hinder an organization in reaching an objective. An arrow pointing to an objective is drawn down the middle of a piece of paper. The factors that will aid the objective's achievement, called the driving forces, are listed on the left side of the arrow. The factors that will hinder its achievement, called the restraining forces, are listed on the right side of the arrow.

**Forced Convection** — Convection occurring due to outside forces such as wind, pumps, or fans.

**Forced Outage** — The shutdown of a generating unit, transmission line, or other facility, in the power industry, for emergency reasons, or a condition in which the generating equipment is unavailable for load due to unanticipated breakdown.

**Forced Vibration** — The oscillation of a system under the action of a forcing function. Typically, forced vibration occurs at the frequency of the exciting force.

**Forcing Frequency** — The frequency at which a shaker vibrates in sinusoidal vibration testing or resonance searching.

**Forcing Function** — A climatic or mechanical environmental input to an item of equipment that affects its design, service life, or ability to function. Synonymous with *environmental condition* and *environmental stress*.

**Forecasting** — Estimating future trends by examining and analyzing available information. Forecasting is the process of estimation in unknown situations.

**Form Factor** — The general shape of a unit or object.

**Form-Fit-Function** — Physical, functional, and performanc characteristics or specifications that uniquely identify a component or device and determine its interchangeability in a system.

| | |
|---|---|
| **Forming** | The first stage of the teaming process. Other stages are:<br>• Storming<br>• Norming<br>• Performing |
| **Fourier Transform Infrared Spectroscopy (FTIS)** | A test where infrared light absorption is used for assessing levels of soot, sulfates, oxidation, nitro-oxidation, glycol, fuel, and water contaminants. |
| **Fourier's Law** | The equation that describes conductive heat transfer through a material where energy transfer equals the product of thermal conductivity, area, and temperature difference. |
| **Fragility** | The maximum load a piece of equipment can stand before failure (malfunction, irreversible loss of performance, or structural damage) occurs. |
| **Fragility Test** | Expensive, but highly useful, dynamic tests of several samples (to account for variations in tolerances, material properties and manufacturing processes) at potentially destructive frequencies, to determine fragility. |
| **Frame – Motor** | Standardized motor mounting and shaft dimensions as established by NEMA or IEC. |
| **Free On Board (FOB)** | The point at which control and title of goods passes to the buyer. After the letters FOB, there is generally a designation of a place where title, control and transportation of goods passes to the buyer. For example, FOB Plant means that the goods pass to the buyer at the seller's plant of origin. |
| **Free Time** | Time during which operational use of the system is not required but the system is operationally ready. |
| **Free Vibration** | Free vibration occurs without forcing, as after a reed is plucked. |
| **Frequency** | The repetition rate of a periodic event, usually expressed in cycles per second (Hz), revolutions per minute (rpm), or multiples of a rotational speed (orders). |
| **Frequency Distribution** | A table that graphically presents a large volume of data so the central tendency (such as the average or mean) and distribution are clearly displayed. |

| | |
|---|---|
| **Frequency Modulation** | The process where the frequency of the carrier is determined by the amplitude of the modulating signal. Frequency modulation produces a component at the carrier frequency, with adjacent components (sideband) at frequencies around the carrier frequency related to the modulating signal. |
| **Frequency Response** | The portion of the frequency spectrum over which a device can be used, within specified limits of amplitude error. |
| **Frequency Spectrum** | A description of the resolution of any electrical signal into its frequency components, giving the amplitude (and sometimes phase) of each component. |
| **Fretting** | A special wear process that occurs at the contact area between two materials under load, and subject to minute relative motion by vibration or some other force. |
| **Fretting Corrosion** | A form of corrosion which can take place when two metals are held in contact and subjected to repeated small sliding, relative motions. Synonymous with *wear oxidation, friction oxidation, chafing,* and *brinelling*. |
| **Friction** | The resisting force encountered at the common boundary between two bodies when, under the action of an external force, one body moves, or tends to move, relative to the surface of the other. |
| **Front-End Loading (FEL)** | The process for conceptual development of processing industry projects. It includes robust planning and design early in a project's lifecycle (the front end of a project) at a time when the ability to influence changes in design is relatively high and the cost to make those changes is relatively low. Synonymous with *pre-project planning (PPP)* and *front-end engineering design (FEED)*. |
| **Fugitive Emissions** | Emissions (air pollutants) released to the air other than those from stacks or vents. They are often due to equipment leaks, evaporative processes, and windblown disturbances. |
| **Full-Flow Filter** | A filter that, under specified conditions, filters all influent flow. |
| **Full-Flow Filtration** | A system of filtration in which the total flow of a circulating fluid system passes through a filter. |

| | |
|---|---|
| **Full-Fluid-Film Lubrication** | The presence of a continuous lubricating film sufficient to completely separate two surfaces, as distinct from boundary lubrication. Full-fluid-film lubrication is normally hydrodynamic lubrication where the oil adheres to the moving part and is drawn into the area between the sliding surfaces forming enough pressure to separate the two surfaces. |
| **Function** | What the owner or user of a physical asset or system wants it to do. |
| **Function Block Diagram** | A top level representation of the major interfaces between a selected system and adjacent systems. |
| **Functional Configuration Audit** | A formal audit to assure that an asset performs to its physical, functional, and performance requirements. This audit is normally done during the final project execution phase of a new or modified asset. |
| **Functional Failure** | A state in which an asset or system is unable to perform a specific function to a level of performance that is acceptable to its user. |
| **Functional Failure Analysis** | The process used to identify and document the system elements, functions, and failure modes that are most important to the disciplines of maintenance and logistics planning, including reliability-centered maintenance. |
| **Functional Failure Modes Effects Analysis** | Top-down analysis of the system or product in which the functional requirements are first totally defined, then an assessment is made of any combination of potential events or conditions that might impair or prevent that function. |
| **Functional Flow Diagrams** | A graphical method for portraying systems in a pictorial manner, illustrating series-parallel relationships, the hierarchy of system functions, and functional interfaces. This is useful in incorporating system operational and maintenance concepts into specific design requirements. |
| **Functional Maintenance Structure** | A type of maintenance organization where the first-line maintenance foreperson is responsible for conducting a specific kind of maintenance, for example, pump maintenance, HVAC maintenance, etc. |
| **Fundamental Frequency** | The number of hertz or cycles per second of the lowest frequency component of a complex, cyclic motion. |

**Fuse**  A piece of metal, connected to a circuit that is to be protected, that melts and interrupts the circuit when excess current flows.

**Fuzzy Logic**  A form of logic used in some expert systems and artificial-intelligence applications in which variables can have degrees of truthfulness or falsehood represented by a range of values between 1 (true) and 0 (false). With fuzzy logic, the outcome of an operation can be expressed as a probability rather than as a certainty. For example, an outcome might be probably true, possibly true, possibly false, or probably false.

# G

**Gain – Control**  The sensitivity or degree of control provided by any manual or automatic adjustment.

**Gain – Signal**  An increase in signal power in transmission from one point to another.

**Gain Sharing**  A reward system that shares the monetary results of productivity gains among owners and employees.

**Galling**  A form of wear in which seizing or tearing of the gear or bearing surface occurs.

**Gantt Chart**  A type of bar chart used in process planning and control to display planned work and finished work in relation to time.

**Gap Analysis**  The comparison of a current condition to the desired state.

**Gas Chromatography**  An analytical method of separating similar vapors by selection adsorption through a specially placed column.

**Gas Tungsten Arc Welding (GTAW)**  Inert gas shielded arc welding using a tungsten electrode. Synonymous with *tungsten inert gas (TIG) welding*.

**Gasohol**  A blend of 10% anhydrous ethanol (ethyl alcohol) and 90% gasoline, by volume. Used as a motor fuel.

**Gatekeeper**  A person who controls access to something.

**Gauge**  A device used as a standard of measurement. An instrument for measuring, indicating, or regulating the capacity, quantity, volume, weight, dimensions, power, amount, properties, etc., of liquids, gases, or solids.

**Gauge Pressure**  The pressure differential above or below atmospheric pressure.

| | |
|---|---|
| **Gauge Repeatability & Reproducibility (GR&R)** | The evaluation of a gauging instrument's accuracy by determining whether the measurements taken with it are repeatable (there is close agreement among a number of consecutive measurements of the output for the same value of the input under the same operating conditions) and reproducible (there is close agreement among repeated measurements of the output for the same value of input made under the same operating conditions over a period of time). |
| **Gaussian Curve** | The charting of a data set in which most of the data points are concentrated around the average (mean), thus forming a bell-shaped curve. Synonymous with *normal distribution*. |
| **Gear Mesh Frequency** | A potential vibration frequency on any machine employing gears. Multiply the number of teeth on a gear times its RPM, then divide by sixty. |
| **Gemba** | A Japanese term for the workplace. |
| **General and Administration Expenses (G&A)** | Management, financial, or other expense incurred by, or allocated to, an organizational unit for the general management and administration of the organization as a whole. |
| **Generated Contamination** | Contamination that is created internally during operation of a hydraulic system due to wear, corrosion, cavitation, or fluid breakdown. |
| **Generation Xers** | Those born between 1960 and 1970. They are very ambitious and independent in nature and strive to balance the competing demands of work, family, and personal life. |
| **Generation Yers** | Those born after 1980. They are technologically savvy with a positive, can-do attitude. Synonymous with *Millennial generation*. |
| **Generator** | Equipment that converts rotational mechanical input power, such as that from a steam turbine, into electricity by using electromagnetic force. |
| **Generic Benchmarking** | Process benchmarking that compares a particular business function or process with other organizations, independent of their industries. |

**Geometric Dimensioning and Tolerancing (GD&T)**  A method to minimize production costs by showing dimensioning and tolerancing, on a drawing, while considering the functions or relationships of part features.

**Glass Fiber Reinforced Concrete**  Concrete-based structural materials used in applications, such as curtain walls, that are manufactured in a molding process where fiberglass is blown in as the concrete is poured.

**Go/No-Go**  The state of a unit or product.

Two parameters are possible:

1. Go – conforms to specifications
2. No-Go – does not conform to specifications

**Goal**  A broad statement describing a desired future condition or achievement without being specific about how much and when.

**Gompertz Model**  A reliability growth model that models reliability values at different stages of development and produces an S-shaped reliability growth curve.

**Grade**  Category or rank given to items that have the same functional use but do not share the same requirements. Note that low grade, unlike low quality, may not be a problem.

**Graduation Mark**  The marks that define the scale intervals on a measuring instrument.

**Graph**  Display or diagram that shows the relationship between activities. Pictorial representation of relative variables. Examples include trend graphs, histograms, control charts, frequency distributions, and scatter diagrams.

**Graphical Evaluation and Review Technique (GERT)**  A network analysis technique, used in project management, that allows probabilistic treatment of both network logic and activity duration estimated.

**Graphical User Interface (GUI)**  A user interface based on graphics, icons, pictures and menus instead of text. Can use a mouse, as well as a keyboard, as an input device.

**Graphite** — A crystalline form of carbon having a laminar structure which is used as a lubricant. It may be of natural or synthetic origin.

**Gravimetric Analysis** — A method of analysis whereby the dry weight of contaminant per unit volume of fluid can be measured showing the degree of contamination in terms of milligrams of contaminant per liter of fluid.

**Graybody** — An object that radiates energy proportional to, but less than, a blackbody at the same temperature.

**Grease** — A lubricant composed of oil or oils thickened with soap or another thickener to a semisolid or solid consistency.

**Green Belt** — An employee of an organization who has been trained (or certified) on the improvement methodology of Six Sigma and will lead a process improvement or quality improvement team.

**Ground** — A large conducting body, such as the earth, used as a common return for an electric circuit and as an arbitrary zero of potential.

**Ground Fault Circuit Interrupter (GFCI)** — A device intended for the protection of personnel, that functions to deenergize a circuit, or portion thereof, within an established period of time when a current to ground exceeds some predetermined value that is less than required to operate the overcurrent protection device of the supply circuit.

**Ground Loop** — A current loop created when a signal source and a signal measurement device are grounded at two separate points on a ground bus through which noise currents flow. These currents generate voltage drops between the two ground connections which cause measurement errors.

**Group Dynamic** — The interaction, or behavior, of individuals within a team meeting.

**Group Leader**  A team member, who may not have any authority over other members, who is appointed on a permanent or rotating basis to represent the team to the next higher reporting level. The leader makes decisions in the absence of a consensus, resolves conflict between team members, and coordinates team efforts. Synonymous with *team leader*.

**Groupthink**  A type of thought exhibited by group members who try to minimize conflict and reach consensus without critically testing, analyzing, and evaluating ideas. Individual creativity, uniqueness, and independent thinking are lost in the pursuit of group cohesiveness. During groupthink, members of the group avoid promoting viewpoints outside the comfort zone of consensus thinking. Tendency of the members of a group to yield to the desire for consensus or unanimity at the cost of considering alternative courses of action. Groupthink is said to be the reason why intelligent and knowledgeable people make disastrous decisions.

**Guideline**  A document that recommends methods to be used to accomplish an objective.

**Gumbel- – Lower**  An extreme minimum value failure distribution, similar to the Weibull distribution. The Gumbel distribution is related to Weibull since Gumbel is a Type I lower extreme value and Weibull is a Type III (lower).

**Gumbel+ – Upper**  An extreme maximum value failure distribution. Gumbel+ (Upper) is a Type I upper extreme value.

# H

**Hard Data** — Measurements such as height, weight, volume, or speed that can be measured on a continuous scale.

**Hard Disk** — A device containing one or more inflexible platters, coated with material in which data can be recorded magnetically, their read/write heads, the head-positioning mechanism, and the spindle motor, enclosed in a sealed case that protects against outside contaminants.

**Hardening** — The process of heating parts to a high temperature and then quenching them in oil, water, air, or solution.

**Hardness – Material** — The resistance of a substance to surface abrasion.

**Hardness – Water** — The scale-forming and lather-inhibiting qualities that water possesses when it has high concentrations of calcium and magnesium ions.

**Hardware** — The physical components of a computer system, including any peripheral equipment such as printers, modems, and mouse devices.

**Harmonic** — Frequency component at a frequency that is an integer multiple of the fundamental frequency.

**Harmonic Distortion** — The presence of harmonics that change an AC waveform from sinusoidal to complex.

**Hash** — A distortion (usually non-harmonic) on a signal.

**Hawthorne Effect** — Generally accepted psychological theory that the behavior of an individual, or a group, will change to meet the expectations of the observer if they are aware their behavior is being observed.

**Hazard** — Any real or potential condition that can cause:

Injury, illness, or death to personnel

Damage to or loss of a system, equipment or property

Damage to the environment

| | |
|---|---|
| **Hazard Analysis (HA)** | An analysis performed to determine how a device, task, location, system, etc., can cause hazards to occur and then how to reduce the severity and probability of occurrence to an acceptable risk level. |
| **HAZard and Operability Study (HAZOP)** | A detailed hazard and operability problem identification process which is carried out by a team. HAZOP deals with the identification of potential deviations from the design intent, examination of their possible causes and assessment of their consequences. |
| **Hazard Rate** | The instantaneous rate of failure for the survivors to time t during the next instant of time. Synonymous with *failure rate*. |
| **Hazardous Material** | Any substance or compound that has the capability of producing adverse health and safety effects on humans or the environment. |
| **Hazardous Waste** | Solid wastes which, because of their quantity, concentration, or physical, chemical, or infectious characteristics, may: <br> 1. Cause or contribute to an increase in mortality and serious, irreversible, or incapacitating illness. <br> 2. Pose a present or potential health hazard when improperly treated, stored, transported, disposed of, or otherwise managed. |
| **Head – Enclosure** | An end closure for the filter case or bowl which contains one or more ports. |
| **Head – Fluid** | The gains, or losses, in pressure caused by gravity and friction as fluid moves through a system. It can be measured in lbs/in2 (psi) or kilopascals (Kpa), but is most commonly expressed in feet of water. |
| **Heat** | Energy associated with the random and chaotic motions of the atomic particles from which matter is composed. |
| **Heat Capacity** | Ability of a material or structure to store heat. It is product of the specific heat and the density of the material. |
| **Heat Exchanger** | A device which transfers heat, through a conducting wall, from one fluid to another. |

| | |
|---|---|
| **Heat Treatment** | Heating and cooling a metal or alloy in such a way as to obtain desired conditions or properties. |
| **Heijunka** | A Japanese term for a system of production smoothing designed to achieve a more even and consistent flow of work. |
| **HEPA Filter** | A high efficiency particulate air filter. |
| **Hertz** | A unit of measure of frequency, usually expressed as cycles per second. |
| **Heuristic Method** | An exploratory method of problem solving in which various types of solutions that may not work are systematically applied and evaluated until a solution is found. |
| **Hidden Failure** | A failure mode that will not become evident to the operating crew under normal circumstances. |
| **Hidden Function** | A function whose failure will not become evident to the operating crew under normal circumstances. |
| **Hidden Plant** | A term describing the production capacity lost to downtime, idle time, startup and transition losses, poor quality, waste and scrap. |
| **Hierarchical** | A group or series classified and arranged in rank order. |
| **High Potential Testing (HiPot Testing)** | A test done to confirm the reliability of an electrical insulation system where a high voltage (twice the operating voltage plus 1000 volts) is applied to cables and motor windings. This is typically a "go/no-go" test. Industry practice calls for HiPot tests on new and rewound motors only. This test stresses the insulation system and can induce premature failures in marginal motors. |
| **High Pressure Pump** | A pump capable of handling flows at significantly higher total dynamic head ratings. Synonymous with *high head pump*. |

**High-Level Language**  A computer language that provides a level of abstraction from the underlying machine language. Statements in a high-level language generally use keywords similar to English and translate into more than one machine-language instruction.

**Highly Accelerated Life Test (HALT)**  A process developed to uncover design defects and weaknesses in electronic and mechanical assemblies using a vibration system combined with rapid high and low temperature changes. The purpose of a HALT is to optimize product reliability by identifying the functional and destructive limits of a product. HALT addresses reliability issues at an early stage in product development.

**High-Pass Filter**  A filter with a transmission band starting at a lower cutoff frequency and extending to infinite frequency.

**Histogram**  A graphic summary of variation in a set of data. The pictorial nature of the histogram allows one to see patterns that are difficult to detect in a simple table of numbers.

**Horizontal Carousels**  Multiple sections of shelving, often called bins, mounted on a revolving track system. They are used for both general inventory and parts storage usage.

**Horsepower (HP)**  Work done over time. The exact definition of one horsepower is 33,000 ft-lb. /minute (745.7 watts). Simply stated, someone lifting 33,000 pounds, one foot, over a period of one minute, is working at the rate of one horsepower.

**Hoshin Kanri**  A Japanese strategic decision-making tool for an organization's executive team that focuses resources on the critical initiatives necessary to accomplish the business objectives of the organization. Hoshin Kanri unifies and aligns resources and establishes clearly measurable targets against which progress toward the key objectives is measured on a regular basis. Synonymous with *Hoshin planning*.

| | |
|---|---|
| **Host Computer** | The main computer, in a mainframe or minicomputer environment, that is the computer to which terminals are connected. In PC-based networks, a computer that provides access to other computers. On the Internet, or other large networks, a server computer that has access to other computers on the network. A host computer provides services such as news, mail, or data, to computers that connect to it. |
| **Hot Work** | Riveting, welding, burning, and other fire or spark producing operations. |
| **House of Quality** | A product planning matrix, somewhat resembling a house, which is developed during quality function deployment and shows the relationship of customer requirements to the means of achieving those requirements. |
| **Housekeeping** | The manufacturing or maintenance activity of identifying and maintaining an orderly environment for preventing errors and contamination in the manufacturing process or work area. |
| **Housing** | A ported enclosure which directs the flow of fluid through the filter element. |
| **Hue** | A characteristic of light at a particular bandwidth that gives a color its name. |
| **Human Factors** | The study of human aspects of systems focusing primarily on cognitive and sensory processes. |
| **Human Factors Analysis** | Analysis covering safety, workmanship, physical design characteristics, and maintainability considerations, and how they impact the operation of equipment and the life cycle of a product. |
| **Human Factors Engineering** | A merging of the branches of engineering and the behavioral sciences which are led principally with the human component in design and operation of human-machine systems. Based on a fundamental knowledge and study of human physical and mental abilities and emotional characteristics. |
| **Human Machine Interface (HMI)** | The boundary at which people make contact with and use machines. When applied to programs and operating systems, it is more widely known as the user interface. |

**Hurdle Rate**  The rate of return, after taxes, that will be used as a criterion for approving projects or investments.

**Hydraulic Fluid**  A fluid serving as the power transmission medium in a hydraulic system. The most commonly used fluids are petroleum oils, synthetic lubricants, oil-water emulsions, and water-glycol mixtures. The principal requirements of a premium hydraulic fluid are:

- Proper viscosity
- High viscosity index
- Anti-wear protection (if needed)
- Good oxidation stability
- Adequate pour point
- Good demulsibility
- Rust inhibition
- Resistance to foaming
- Compatibility with seal materials

Anti-wear oils are frequently used in compact, high-pressure, and capacity pumps that require extra lubrication protection.

**Hydraulic Oil**  Oil used in hydraulic systems.

**Hydraulics**  The engineering science pertaining to liquid pressure and flow.

**Hydrocarbon**  Compounds containing only carbon and hydrogen. Petroleum consists chiefly of hydrocarbons.

**Hydrodynamic Lubrication**  A system of lubrication in which the shape and relative motion of the sliding surfaces causes the formation of a fluid film having sufficient pressure to separate the surfaces.

**Hydrofinishing**  A process for treating raw extracted base stocks with hydrogen to saturate them for improved stability.

**Hydrolysis**  A breakdown process that occurs in anhydrous hydraulic fluids as a result of heat, water, and metal catalysts – iron, steel, copper, etc.

**Hydrometer**  An instrument for determining either the specific gravity of a liquid, or the API gravity.

**Hydrostatic Lubrication**  A system of lubrication in which the lubricant is supplied under sufficient external pressure to separate the opposing surfaces by a fluid film.

**Hyperlink**  A connection between an element in a hypertext document (a word, phrase, symbol, or image) and a different element in the document (another document, file, or script). The user activates the link by clicking on the linked element which is usually underlined, or a different color from the rest of the document, to indicate that the element is linked.

**HyperText Markup Language (HTML)**  The markup language used for documents on the World Wide Web.

**Hyperthermia**  Heating so excessive that it can damage or kill plant or animal cells.

**Hypoid Gear Lubricant**  A gear lubricant, having extreme pressure characteristics, for use with a hypoid type of gear, as in the differential of an automobile.

**Hypothesis Testing**  The use of statistics to determine the probability that a given hypothesis is true.

The usual process of hypothesis testing consists of four steps:

1. Formulating the null hypothesis and the alternative hypothesis.
2. Identifying a test statistic that can be used to assess the truth of the null hypothesis.
3. Computing the probability (P-value) that a test statistic at least as significant as the one observed would be obtained assuming that the null hypothesis was true.
4. Comparing the P-value to an acceptable significance value (sometimes called an alpha value). If the observed effect is statistically significant, the null hypothesis is ruled out and the alternative hypothesis is deemed valid.

**Hysteresis**      That portion of a measuring system's response where a change in input does not produce a change in output. Mathematically, it is a non-uniqueness in the relationship between two variables as a parameter increases or decreases. Synonymous with *dead band*.

# I

**Ideal Gas Law**     The equation that describes the relationship among pressure, temperature and volume of a gas.

**Idle Time**     The amount of time an asset is idle or waiting to run. It is the sum of the times when there is no demand, no feedstock or raw material, or down for administrative reasons (i.e., times when the asset is not scheduled for production).

**Illuminance**     The density of luminous flux on a surface. Measured in the SI system by lux.

**Image Analyzer**     A sophisticated microscopic system involving a microscope, a television camera, a dedicated computer, and a viewing monitor.

**Image Processing**     The analysis, manipulation, storage and display of graphical images from sources such as photographs, drawings and video.

**Imbalance**     The unequal distribution of weight or mass on a rotor. A shaft condition such that the mass and shaft geometric center lines do not coincide.

**Immiscible**     Incapable of being mixed without separation of phases. Water and petroleum oil are immiscible under most conditions, although they can be made miscible with the addition of an emulsifier.

**Impact**     The single, instantaneous stroke of a body in motion against another, either in motion or at rest.

**Impact – Project**     The effect of a change in cost or schedule on a project.

**Impact Testing**     Testing an object's ability to resist high-rate loading. An impact test is a test for determining the energy absorbed in fracturing a test piece at high velocity.

**Impedance – Electrical**     The total resistance to flow of an alternating current, generally expressed in ohms. It is a combination of resistance and reactance.

| | |
|---|---|
| **Impedance – Mechanical** | The mechanical properties of a machine system (mass, stiffness, damping) that determine the response to the periodic forcing functions. |
| **Impeller** | A rotating disk with a set of vanes coupled to the engine or drive shaft that produces centrifugal force within the pump casing of a centrifugal pump. |
| **Imperfection** | A quality characteristic's departure from its intended level, or state, without any association to conformance to specification requirements, or to the usability of a product or service. |
| **Impregnation** | The coating and binding, in electric motors, that holds components together, fills in the air space, and provides protection against contaminants. It is applied in a fluid form and hardened. |
| **Improvement** | The positive effect of a process change effort. |
| **Impulse** | The integral of force over a time interval. |
| **Inactive Inventory** | Inventory that shows no usage history, usually twelve months or longer. |
| **Inactive Stocked MRO Inventory Value** | The current book value of maintenance, operating, and repair (MRO) materials held in stock at the plant site which shows no usage history (including consignment and vendor-managed stores). It includes the value of inactive stocked MRO materials in all stored locations, including remote stock locations, whether or not that material is officially accounted for in the inventory asset accounts, or in an allocated portion of pooled spares. |
| **Incandescent** | Emitting visible radiation as a result of heating. |
| **Incipiency** | Progressive performance deterioration which can be measured using instruments. |
| **In-Control Process** | A process in which the statistical measure being evaluated is in a state of statistical control. It indicates, within limits, a predictable and stable process. |

**Incremental Cost**  The change in total cost that arises when the quantity produced changes by one unit. It is the cost of producing one more unit of a product. Synonymous with *marginal cost*.

**Indenture Level**  A designation which identifies an item's relative complexity as an assembly or function. In a system, the first indenture level is the system. Examples of lower indenture levels could be system segments (level 2), prime items (level 3), subsystems (level 4), components (level 5), subassemblies (level 6), and parts (level 7). Synonymous with *asset hierarchy*.

**Indicators**  Established measures used to determine how well an organization is meeting its customers' needs, as well as other operational and financial performance expectations.

**Individual Tool Kit**  An individual-user container used to store tools or equipment, enabling effective/efficient control of tools and ease of inventory.

**Indoor Air Quality**  A measure of indoor pollution from sources that release gases or particles into the air. Inadequate ventilation can increase indoor pollutant levels by not bringing in enough outdoor air to dilute emissions from indoor sources and by not carrying indoor air pollutants out from the building or facility. High temperature and humidity levels can also increase concentrations of some pollutants.

There are three basic strategies to improve indoor air quality:

1. Source control
2. Improve ventilation
3. Install air cleaners or filters

**Induced Environments**  Conditions generated by operating some equipment, as opposed to natural environments.

**Inductance**  The characteristic of an electric circuit by which varying current in it produces a varying magnetic field causing voltages in the same or a nearby circuit.

| | |
|---|---|
| **Induction Motor** | An alternating current (AC) motor in which the primary winding on one member (usually the stator) is connected to the power source and a secondary winding, or a squirrel-cage secondary winding, on the other member (usually the rotor) carries the induced current. There is no physical electrical connection to the secondary winding, its current is induced. |
| **Inductor** | A device that stores electrical energy in a magnetic field. |
| **Inertance** | A measure of the pressure gradient in a fluid required to cause a change in flow-rate with time. |
| **Inertance – Acoustic** | The effect of inertia in an acoustic system which impedes the transmission of sound through the system. |
| **Infant Mortality** | Failures that occur prematurely. Such failures can occur because of inadequate design, inferior material, poor workmanship, improper installation, or quality problems in any work that was done prior to an asset entering service. |
| **Inferential Statistics** | The techniques for reaching conclusions about populations on the basis of analysis of a data from a sample. |
| **Influent** | The fluid entering a component. |
| **Infrared (IR)** | Radiant energy, beyond the color red, of wavelengths from the red visible (0.75 μm) to about 300 μm, between the visible and microwave regions of the electromagnetic spectrum. |
| **Infrared Spectra** | A graph of infrared energy absorbed at various frequencies in the additive region of the infrared spectrum. The current sample, the reference oil, and the previous samples are usually compared. |
| **Infrared Spectroscopy** | An analytical method using infrared absorption for assessing the properties of used oil and certain contaminants suspended therein. |
| **Infrared Thermal Imager** | Instrument, or system, that converts incoming infrared radiant energy from a target surface to a thermal map, or thermogram, on which color hues or gray shades can be related to the temperature distribution on the surface. |

| | |
|---|---|
| **Infrared Thermography** | A monitoring technique that uses special instruments, such as an infrared camera, to detect, identify, and measure the heat energy objects radiate in proportion to their temperature and emissivity. Synonymous with *infrared monitoring*. |
| **Ingested Contaminant** | An environmental contaminant that ingresses due to the action of the system or machine. |
| **Ingression Level** | Particles added per unit of circulating fluid volume. |
| **Inherent** | Existing in someone, or something, as a natural and inseparable quality or characteristic. |
| **Inherent Availability (Ai)** | The probability that, when used under stated conditions in an ideal support environment without consideration for preventive action, a system will operate satisfactorily at any time. Inherent availability excludes whatever ready time, preventive maintenance downtime, supply downtime, and administrative downtime may be required. |
| **Inherent Reliability** | The reliability that is designed into an item. |
| **Inhibitor** | Any substance that slows or prevents such chemical reactions as corrosion or oxidation. |
| **Initial Capability** | The level of performance that a physical asset or system is capable of achieving at the moment it enters service. |
| **Injury and Illness Incident Rate** | The number of recordable injuries and illnesses occurring among a given number of full-time workers (usually 100 full-time workers) over a given period of time (usually one year). Organizations must report these rates to OSHA using Form 300. |
| **In-Line Filter** | A filter assembly in which the inlet, outlet, and filter element axes are in a straight line. |
| **In-Plant Defect Rate** | The fallout rate, expressed as parts per million (ppm), of all components in manufacturing and assembly that fail quality tests at any point in the production process. |
| **Input – Computer** | Information entered into a computer or program for processing, as from a keyboard or from a file stored on a disc drive. |

| | |
|---|---|
| **Input – Process** | Any item, whether internal or external to the project, which is required by a process before that process proceeds. May be an output from a predecessor process. |
| **Input Control Signal** | A signal originating in a control sensor. Sometimes selected, or averaged, between several sensors. |
| **Input Impedance** | The shunt resistance and capacitance (or inductance) as measured at the input terminals, not including effects of input bias or offset currents. |
| **Input Signal** | A signal applied to a device, element, or system. |
| **In-Service Inspection (ISI)** | A visual inspection aimed at discovering asset faults before they result in an operational failure. The inspection is routinely performed on hydraulic and piping systems. |
| **In-Situ** | A term meaning "at site". Usually used in the context of performing work on an item without moving the item to a workshop. |
| **Insolubles** | Particles of carbon, or agglomerates of carbon and other material. Indicates deposition or dispersant drop-out in an engine. Not serious in a compressor or gearbox unless there has been a rapid increase in these particles. |
| **Insourcing** | Using in-house resources to deliver a service or material, or to fabricate equipment or components. |
| **Insourcing Maintenance** | The process of moving maintenance activities performed by outside contractors in-house to be performed by company employees. Antonym of outsourcing maintenance. |
| **Inspection** | Measuring, examining, testing, and gauging one or more characteristics of a product or service and comparing the results with specified requirements to determine whether conformity is achieved for each characteristic. |
| **Inspection Cost** | The cost associated with inspecting a product to ensure it meets the internal or external customer's needs and requirements. |
| **Instantaneous Field Of View (IFOV)** | The specification of a system detailing the smallest area that can be accurately seen at a given instant. Synonymous with *special resolution*. |

| | |
|---|---|
| **Instrument** | The term for any item of electrical or electronic equipment designed to carry out a specific function or set of functions. A device for measuring the value of an observable attribute. |
| **Instrumentation** | Devices, interconnections, and systems used to observe, measure, and/or provide data as to what is, or has occurred, in order to evaluate or control physical phenomenon or processes. |
| **Insulation** | A non-conductive material used to separate conducting materials in a circuit. Synonymous with *insulator*. |
| **Insulation Class** | The classification of electric motor insulation by the temperature ranges at which it can operate for a sustained period of time without degradation. |
| **Insulation Resistance Profile (IRP)** | A graph of resistance versus time which provides useful information in addition to the standard insulation resistance measurement, particularly in ranges greater than 5,000 megaohms. Synonymous with *polarization index profile*. |
| **Integrated** | Interrelated, interconnected, interlocked, or meshed components blended and unified into a functioning or unified whole. |
| **Integrated Circuit (IC)** | A small, complete circuit of interconnected semiconductor devices such as transistors, capacitors, and resistors printed on a single silicon chip. |
| **Intensifier** | A device which converts low pressure fluid power into higher pressure fluid power. |
| **Intensity** | The severity of a vibration or shock. Nearly the same meaning as Amplitude, defined earlier, but less precise, lacking units. |
| **Interaction** | The effect of one part, element, subsystem, or system on another. |
| **Interfacial Tension** | The energy per unit area present at the boundary of two immiscible liquids. It is usually expressed in dynes/cm (ASTM Designation D 971). |
| **Interlock** | A device that prevents a machine from initiating an operation until a condition, or set of conditions, is fulfilled. |

| | |
|---|---|
| **Intermediate Customers** | Organizations or individuals who operate as distributors, brokers, or dealers between the supplier and the consumer/end user. |
| **Intermittent Failure** | Failure, for a limited period of time, followed by the item's recovery of its ability to perform within specified limits without any remedial action. |
| **Internal Benchmarking** | Benchmarking that is performed within an organization by comparing similar business units or business processes. |
| **Internal Customer** | The recipient (person or department), within an organization, of another person's or department's output (product, service or information). See *external customer*. |
| **Internal Rate of Return (IRR)** | The rate of compound interest at which a company's outstanding investment is repaid by proceeds of a project. A discount rate that causes net present value to equal zero. |
| **International Organization for Standardization (ISO)** | The world's largest developer and publisher of International Standards. The organization is a network of the National Standards Institutes of 159 countries, with a Central Secretariat in Geneva, Switzerland that coordinates the system. |
| **International System of Units (SI)** | An internationally accepted system of measurement based upon nine precisely defined metric units: <br> 1. Kilogram (mass) <br> 2. Meter (length) <br> 3. Second (time) <br> 4. Ampere (current) <br> 5. Kelvin (temperature) <br> 6. Candela (light intensity) <br> 7. Mole (molecular atomic weight) <br> 8. Radian (plane angle) <br> 9. Steradian (solid angle) |
| **Internet** | The worldwide collection of networks and gateways that use the TCP/IP suite of protocols to communicate with one another. |
| **Interoperability** | A property referring to the ability of diverse systems and organizations to work together, or interoperate. |

| | |
|---|---|
| **Interrelationship Digraph** | A management tool that depicts the relationship among factors in a complex situation. Synonymous with *relations diagram*. |
| **Interruptive Task** | Any maintenance task which interrupts the normal operation of a machine, system, or asset. |
| **Intervention** | The action of a team facilitator, when interrupting a discussion, to state observations about group dynamics or the team process. |
| **Intranet** | A private network based on Internet protocols such as TCP/IP but designed for information management within a company or organization. |
| **Intrinsic Availability** | The probability that a system is operating satisfactorily at any point after the start of operation, when operated under stated conditions, when the only times considered are operating time, and active repair time. |
| **Intrusive Task** | Actions that require process or asset interruption, equipment shutdown, tagout, entry or disassembly. |
| **Inventory** | The value of raw materials, products in process, and finished products required for plant operation. Also includes other supplies, i.e., for maintenance, supplies and spare parts. |
| **Inventory Control** | Managing the acquisition, receipt, storing, and issuance of materials and spare parts. Managing the stores inventory effectively. |
| **Inventory Line Items** | Number of line items in the local storeroom (stockroom, crib) that have a stock level of one or more units that are used to support Maintenance Repair and Operation (MRO). Does not include line items that are held off-site by an outside supplier organization. |
| **Inventory Stock Record** | The data entry describing the part that is inventoried and is represented by the stock keeping unit or SKU. The inventory stock record should not be confused with the quantity in stock. |
| **Inventory Turnover** | Ratio of the value of materials and parts issued annually to the value of materials and parts on-hand, expressed as a percentage. |

| | |
|---|---|
| **Inverter** | An electronic device that converts fixed frequency and voltage to variable frequency and voltage. It enables the user to electrically adjust the speed of an AC motor. |
| **Irradiance** | The power of electromagnetic radiant energy incident on, or radiated from, the surface of a given unit area. |
| **Ishikawa Analysis** | A graphical technique that can be used to identify and arrange the causes of an event, problem, or outcome. The hierarchical relationships between the causes according to their level of importance or detail are illustrated on branches. Synonymous with *cause-and-effect* and *fishbone analysis*. |
| **Ishikawa Diagram** | The diagram resulting from an Ishikawa analysis. Synonymous with *cause-and-effect* and *fishbone diagram*. |
| **ISO 14000** | An environmental management standard related to what organizations do that affects their physical surroundings. In the process of being made compatible with ISO 9000. |
| **ISO 4021** | The international standard which describes the method for extracting fluid samples from a main flow line in such a manner that the particulate contaminant in the sample is representative of the fluid flowing at the point of sampling. |
| **ISO 9000 Series Standards** | A set of international standards on quality management and quality assurance developed to help companies effectively document the quality system elements to be implemented to maintain an efficient quality system. The standards, initially published in 1987, are not specific to any particular industry, product, or service. |

The standards underwent major revision in 2000 and now include:

- ISO 9000:2000 (definitions)
- ISO 9001:2000 (requirements)
- ISO 9004:2000 (continuous improvement)

| | |
|---|---|
| **ISO Viscosity Grade** | A number indicating the nominal viscosity of an industrial fluid lubricant at 40°C (104°F) as defined by ASTM Standard Viscosity System for Industrial Fluid Lubricants D 2422. |

**Isotherm**  A locus or pattern superimposed on a thermogram or line scan that includes all points that have the same apparent temperature.

**Isotropy**  A condition in which significant medium properties, such as velocity, are the same in all directions.

**Issue**  An important question that is in dispute and must be settled.

**Item**  A non-specific term used to denote any product, including systems, materials, parts, subassemblies, sets, accessories, etc.

**Item Readiness Review**  The final review, held jointly between customer and system managers, to ascertain that an item meets all technical requirements, is fully documented, and is ready for turnover to the customer, or user in operations after a major modification or repair.

**Iterative Closed Loop**  A control system that pre-calculates drive signals but then modifies those signals based upon resulting motion, in order to better match measured with desired motions. Evaluation and modifications take place after each excitation, repeating until the match is acceptable.

# J

**Jackscrew** — A device used for leveling and/or positioning of a piece of equipment, such as a motor. These devices are adjustable screws that fit on the base or the equipment.

**Jidoka** — A Japanese term for a method of autonomous control involving the adding of intelligent features to machines to start or stop operations as control parameters are reached and to signal operators when necessary. Synonymous with *autonomation*.

**Jisha Kanri** — A Japanese term for self-management or voluntary participation.

**Job** — A work assignment, task or related series of tasks, position classification, or quantity of work.

**Job Characteristic** — An attribute of a particular work function related to skill, experience, responsibility, effort, or working conditions.

**Job Description** — An established, written, summary statement of a position which describes its functional requirements. It may include such specifics as equipment or tools used, physical and mental skills required, training, working conditions, duties, responsibilities, and designation of supervisor. Synonymous with *position description*.

**Job Plan** — Instructional and supporting information (hardcopy and/or electronic) provided to the work performer. It contains job-specific requirements such as task descriptions sequenced in steps, job-specific and safety permits, procedures, drawings, and materials.

**Job Task Analysis** — An analysis of the various elements pertaining to skill requirements, training and experience, mental and physical demands, working conditions, hazards exposure, requirements and responsibilities of performance for a particular job.

**Job Tool Card** — A record kept by a tool room to keep track of tools issued to jobs rather than to individual technicians.

**Joule** — A unit of measurement of energy equal to a watt-second.

**Journal – Document** — A chronological documentation of activities in a book. Synonymous with *log book* and *log*.

**Journal – Mechanical** — That part of a shaft or axle that rotates, or angularly oscillates, in or against a bearing, or about which a bearing rotates or angularly oscillates.

**Journal Bearing** — A sliding type of bearing having either rotating or oscillatory motion and in conjunction with which a journal operates.

**Journeyman** — A fully qualified trade craftsman who has successfully completed an apprentice program.

**Junction – Analysis** — An object used in reliability block diagrams to join multiple connectors together, such as in parallel or redundant configurations. Junctions can be set in series, parallel operating, or standby operation.

**Juran Trilogy** — Three managerial processes identified by J.M. Juran for use in managing for quality:
1. quality planning
2. quality control
3. quality improvement

**Just-In-Time (JIT)** — A manufacturing process technique in which there is little or no material inventory kept on hand. Everything arrives just-in-time to be used in next process step.

# K

**Kaikaku**     A Japanese term for radical improvement of an activity to eliminate muda (waste). Synonymous with *breakthrough kaizen*.

**Kaizen**     A Japanese term that means gradual, unending improvement by doing little things better along with setting and achieving increasingly higher standards.

**Kaizen Event**     A concentrated effort, typically spanning one to five days, in which a team plans and implements a major process change to improvement in performance. Participants generally represent various functions and perspectives and may include non-plant personnel.

**Kanban**     A Japanese term for a tool that helps to maintain an orderly and efficient flow of materials throughout the entire manufacturing process. It is a material requirement planning technique used in Just-In-Time (JIT) inventory systems in which work-centers signal with a card when they wish to withdraw parts from feeding operations or supply bins. Kanban means a visible record (such as a card, label, or sign) in Japanese.

**Kanban Signal**     A method of signaling suppliers, or upstream production operations, when it is time to replenish limited stocks of components or subassemblies in a just-in-time system.

**Kano Analysis**     A method of classifying and prioritizing customer requirements to reveal which are most important to satisfaction and which may be limited without damaging overall satisfaction.

| | |
|---|---|
| **Karl Fischer Reagent Method** | The standard laboratory test to measure the water content of mineral base fluids. In this method, water reacts quantitatively with the Karl Fischer reagent. This reagent is a mixture of iodine, sulfur dioxide, pyridine, and methanol. When excess iodine exists, electric current can pass between two platinum electrodes or plates. The water in the sample reacts with the iodine. When the water is no longer free to react with iodine, an excess of iodine depolarizes the electrodes signaling the end of the test. ASTMD-1744-64. |
| **Karoshi** | A Japanese term for death from overwork. |
| **Kelvin** | Absolute temperature scale related to the Celsius (or centigrade) scale. |
| **Key Performance Indicator (KPI)** | A vital indicator which is used to evaluate process performance. KPIs are important management tools that measure business performance, including maintenance. |
| **Key Process** | A major, system level process that supports the mission and satisfies major consumer requirements. |
| **Keyword** | A characteristic word, phrase, or code that is stored in a key field and is used to conduct sorting or searching operations on records in a database. |
| **Kinematic Viscosity** | The quotient of the absolute viscosity in centipoises divided by the specific gravity of a fluid, both at the same temperature. The units of kinematic viscosity are stoke and centistokes (1/100 of a stoke). |
| **Kirchhoff's Law** | The equation that states, for an opaque object, radiant energy absorbed equals radiant energy emitted. |
| **KISS Model** | A pragmatic philosophy of conducting business in which the objective is to keep things such as procedures, reports, and any other aspect of work, as simple as possible to get the job done. The acronym humorously describes the basic premise of simplicity which is, "Keep It Simple and Specific." |

| | |
|---|---|
| **Kitting** | A process of assembling all the parts and tools for a job task into a box or bin. This procedure eliminates time consuming trips from one parts bin, tool crib, or supply center to another to get the necessary material. The kit is delivered, in advance, to the work site before the work starts. |
| **Knowledge Base** | The fundamental body of knowledge available to an organization, including the knowledge in people's heads, supported by the organization's collections of information and data. An organization may also build subject-specific knowledge bases to collate information on key topics. Synonymous with *database of information*. |
| **Knowledge Management** | The deliberate and systematic coordination of an organization's people, technology, processes, and organizational structure in order to add value through reuse and innovation. This value is achieved through the promotion of creating, sharing, and applying knowledge, as well as through the feeding of valuable lessons learned and best practices. |
| **Knowledge Management System** | A system that is designed to capture the explicit knowledge of a company's employees, contractors, and other people who work on-site, on either a permanent or temporary basis. |
| **Kobetsu-Kaizaen** | A Japanese term referring to an individual improvement to further improve production systems. It involves efforts to select a model piece of equipment and challenge the target of zero losses through project team activities. |

# L

**Labor** — The physical effort a person has to expend to repair, inspect, or deal with a problem.

**Labor Availability** — The percentage of time that the maintenance crew is available to perform productive work during a scheduled working period.

**Labor Costs** — The total cost of labor including benefits.

**Labor Productivity** — A partial productivity measure. The rate of output of a worker, or group of workers, per unit of time compared to an established standard or rate of output. Expressed as output per unit of time or output per labor hour.

**Labor Turnover Rate** — A measure of a plant's ability to retain workers. Expressed as a percentage of the production workforce that annually departs, regardless of reason. High turnover rates often indicate employee dissatisfaction with either working conditions or compensation.

**Lack of Fusion** — Discontinuity due to lack of union between weld metal and parent metal or between successive weld beads. Synonymous with *incomplete penetration*.

**Lacquer** — A deposit resulting from the oxidation and polymerization of fuels and lubricants when exposed to high temperatures. Similar to, but harder, than varnish.

**Ladder Logic** — A symbol system used to illustrate the functions of a control circuit schematically. Power lines form the sides of the ladder-like structure, with program elements arranged to form the rungs.

**Lagging Indicator** — An indicator that measures performance after the business or process results start to follow a particular pattern or trend. Lagging indicators confirm long-term trends, but do not predict them.

**Lambertian** — Having a surface that emits uniformly in all directions. A blackbody is a Lambertian source.

| | |
|---|---|
| **Laminar Particles** | The particles generated in rolling element bearings which have been flattened out by a rolling contact. |
| **Laser** | Short for Light Amplification by Stimulated Emission of Radiation. A laser produces a highly monochromatic and coherent beam of radiation. |
| **Last-In-First-Out (LIFO)** | An inventory valuation method in which costs are transferred in reverse chronological order. |
| **Late-Finish Date** | In the critical path method, the latest possible point in time that a schedule activity may be completed without violating schedule constraint or delaying the project completion date. |
| **Late-Start Date** | In the critical path method, the latest possible point in time that a schedule activity may begin without violating schedule constraint or delaying the project completion date. |
| **Latent Defect** | A flaw, in a part, assembly, and/or workmanship that is not immediately apparent visually or by testing methods, yet can result in failure. |
| **Latent Heat** | The amount of energy released or absorbed by a substance during a change of state. |
| **Latent Heat of Fusion** | The amount of energy released or absorbed by a substance during a phase change from a solid to a liquid (melting). |
| **Latent Heat of Vaporization** | The amount of energy released, or absorbed, by a chemical substance during a phase change from a liquid to a gas (vaporization or boiling). |
| **Lead Time** | The time required to wait for a product, service, material, or after ordering or making a request for such things. |
| **Leader** | An individual who is recognized by others as a person they will follow. |
| **Leadership** | An essential part of an improvement effort. Organization leaders must establish a vision, communicate that vision to those in the organization, and provide direction, resources and knowledge necessary to achieve goals and accomplish the vision. |

| | |
|---|---|
| **Leading Indicator** | An indicator that measures performance before the process results start to follow a particular pattern or trend. Leading indicators can be used to predict changes and trends. |
| **Leak** | An opening that allows the passage of a fluid. |
| **Leak Testing** | A nondestructive testing method for detecting, locating, or measuring leaks or leakage in pressurized or evacuated systems or components. |
| **Lean** | A practice that considers the expenditure of resources for any goal other than the creation of value for the end customer to be wasteful, and thus a target for elimination. |
| **Lean Initiatives** | Business improvement initiatives that are designed to remove waste from the business processes. The waste may include materials, time, scrap, poor quality, no value add tasks, buffers, work-in-progress. |
| **Lean Maintenance** | The practice of applying the principles of lean to a maintenance organization. |
| **Lean Manufacturing** | An initiative focused on eliminating all waste in manufacturing processes. Principles of lean include zero waiting time, zero inventory, scheduling (internal customer pull instead of push system), batch size optimization, line balancing and cutting actual process times. |
| **Learning Curve** | A graphical representation of the common sense principle that the more one does something, the better one gets at it. A learning curve shows the rate of improvement in performing a task as a function of time, or the rate of change in cost or efficiency as a function of cumulative output. |
| **Learning Organization** | An organization that possesses the practices, systems, and culture that actively promotes sharing of experiences and lessons learned to encourage quality performance and continuous improvement. |
| **Lessons Learned** | The knowledge that results from a postmortem or after-the-fact analysis of a project or application of a new technique. |

**Lessons-Learned Review** — An audit or evaluation conducted immediately upon completion of a project or activity to learn from the successes and failures recently experienced. The results of the review are documented for use by others as a reference guide in the future.

**Level of Repair Analysis (LORA)** — An analysis conducted to determine whether an item should be repaired at a given level of repair or maintenance. The analysis considers such variables as cost of repair, the unit cost of the item, reliability of the item, and turn-around time for each potential level of maintenance.

**Level of Service** — The degree of maintenance performed to meet desired levels of equipment performance. A high level ensures little chance of failure while a low level meets minimum requirements, risking breakdowns on less critical equipment.

**Life Cycle** — The entire useful life of a product or service, usually divided into sequential phases, which include design, development, build, operate, maintain, and disposal (or termination).

**Life Cycle Cost (LCC)** — All costs associated with the items of life cycle including: design, development, build, operate, maintain, disposal. It is a total cost of ownership for the life of the asset. Synonymous with *total cost of ownership*.

**Life Cycle History** — A time history of events and conditions associated with an item of equipment from its release from manufacturing to its removal from service.

The life cycle should include the various phases that an item will encounter in its life, such as:

- Handling, shipping and storage prior to use.
- Mission profiles while in use.
- Phases between missions, such as standby time or storage, transfer to and from repair sites, and alternate locations.
- Geographical locations of expected deployment.

**Light Obscuration** — The degree of light blockage as reflected in the transmitted light impinging on a photodiode.

**Light-Emitting Diode (LED)** — A solid-state device that radiates in the visible region. Used in alphanumeric displays and as indicator or wiring lights.

**Limit Switch** — An electromechanical device positioned to be actuated when a certain motion limit occurs, thereby deactivating the actuator causing the motion.

**Line Chart** — A type of graph created by connecting a series of data points together with a line. It is an extension of a scatter graph and is often used to visualize a trend in data over intervals of time, thus the line is often drawn chronologically. Synonymous with *line graph*.

**Line Items in Inventory** — The number of different items in inventory, each with its own unique description and stock number.

**Linear** — Said of any device or motion where the effect is exactly proportional to the cause.

**Linear Programming** — A management technique applied to problems in which a linear function of a number of variables is subject to a number of constraints in the form of linear inequalities.

**Linear System** — A system where its magnitude of response is directly proportional to its magnitude of excitation, for every part of the system.

**Linearity** — Refers to the closeness of a calibration curve to a straight line, in alignment. Having output directly proportional to input.

**Liquid** — Any substance that flows readily, or changes in response to the smallest influence.

**Liquid Crystal Display (LCD)** — A type of display that uses a liquid compound having a polar molecular structure, sandwiched between two transparent electrodes. When an electric field is applied, the molecules align with the field forming a crystalline arrangement that polarizes the light passing through it. A polarized filter laminated over the electrodes blocks polarized light. In this way, a grid of electrodes can selectively "turn on" a cell, or pixel, containing the liquid crystal material, turning it dark.

**Load – Electrical** — Anything in an electrical circuit that, when the circuit is turned on, draws power from that circuit.

| | |
|---|---|
| **Load Cell** | A transducer for the measurement of force or weight. Action is based on strain gages mounted within the cell on a force beam. |
| **Load Factor** | The average load carried by an engine, machine, or plant expressed as a percentage of its maximum capacity. |
| **Load-Carrying Capacity – Lubrication** | The property of a lubricant to form a film on the lubricated surface which resists rupture under given load conditions. Expressed as maximum load the lubricated system can support without failure or excessive wear. |
| **Loading Capacity** | The safe working capacity determined in accordance with the applicable code for allowable loads and working stresses. |
| **Local Area Network (LAN)** | A group of computers, and other devices, dispersed over a relatively limited area and connected by a communications link that enables any device to interact with any other on the network. |
| **Lock Nut** | A nut used in combination with a lock washer to hold a bearing in place on a shaft. |
| **Lock Washer** | A washer with tongue and prongs to hold a lock nut in place. |
| **Log** | A document used to record and describe or denote selected items identified during execution of a process or activity. |
| **Logic Tree Analysis (LTA)** | An analytical method that uses deductive logic to guide thought process used to draw correct conclusions. LTAs are used to lay out the logic of systems to aid in human decision making, and are used extensively in root cause failure analysis (RCFA). Sometimes used interchangeable with fault tree analysis, but the latter generally applies probabilities to determine causes. |
| **Logistics Support** | The materials and services required to operate, maintain, and repair a system. Logistics support includes the identification, selection, procurement, scheduling, stocking, and distribution of spares, repair parts, facilities, support equipment, trainers, technical publications, contractor engineering and technical services, and personnel training necessary to provide the capabilities required to keep the system in functioning status. |

| | |
|---|---|
| **Logistics Support Analysis** | A methodology for determining the type and quantity of logistic support required for a system over its entire lifecycle. Used to determine the cost effectiveness of asset based solutions. |
| **Lognormal Distribution** | A single-tailed probability distribution of any random variable whose logarithm is normally distributed. If $X$ is a random variable with a normal distribution, then $Y = \exp(X)$ has a log-normal distribution; likewise, if $Y$ is log-normally distributed, then $\log(Y)$ is normally distributed. The lognormal distribution is commonly used to model the lives of units whose failure modes are of a fatigue-stress nature. The lognormal distribution can have widespread application, and is a good companion to the Weibull distribution. |
| **Lost Opportunity Event (LOE)** | Any event that causes production to be curtailed (either partially or completely) for any reason. |
| **Lost Time Incident** | An incident that results in personnel being unable to perform their work for some period of time. A now-outdated term still in common use is *lost time accident*. |
| **Lot** | A defined quantity of product accumulated under conditions considered uniform for sampling purposes. |
| **Lot Quality** | The value of percentage defective, or defects per hundred, units in a lot. |
| **Lot Size** | The number of units in a lot. Referred to as N. |
| **Lot Tolerance Percentage Defective** | The poorest quality, expressed in percentage defective, in an individual lot that should be accepted. |
| **Loudness** | The human ranking of an auditory sensation, usually in terms ranging from soft to loud. Expressed in sones (not decibels). |
| **Lower Control Limit (LCL)** | The control limit for points below the central line in a control chart. |
| **Lower Explosive Limit (LEL)** | The lowest concentration (percentage) of a gas or vapor in air capable of producing a flash of fire in the presence of an ignition source (arc, flame, heat). Concentrations lower than LEL are too lean to burn. Synonymous with *lower flammable limit (LFL)*. |

| | |
|---|---|
| **Lubricant** | Any substance interposed between two surfaces in relative motion for the purpose of reducing the friction and/or wear between them. |
| **Lubricate** | The adding of a type of lubrication, typically grease or oil, into a compartment, surface, or onto an exposed component. |
| **Lubrication Task** | A time or condition directed action involving addition or exchange of lubricant (such as grease or oil). |
| **Lubricity** | The ability of an oil or grease to lubricate. Synonymous with *film strength*. |
| **Luminance** | The ratio of a surface's luminous intensity in a given direction to a unit of projected area. |
| **Luminosity** | The luminous efficiency of radiant energy. |
| **Lux (lx)** | The SI unit of measure for illuminance. Equivalent to lumens per square meter. Synonymous with *meter-candle*. |

# M

**Machine Code** — The ultimate result of the compilation of assembly language, or any high-level language such as C or Pascal. A sequence of 1s and 0s that are loaded and executed by a microprocessor.

**Machine Capability Index (Cmk)** — An index derived from the observations from an uninterrupted production run on a machine.

**Machine Vision** — An automated optical or video system for acquiring, processing, and analyzing images to evaluate a test object or to provide information for human interface and interpretation. Typical applications are to guide robot or automated systems for quality control.

**Machinery Information Management Open Systems Alliance (MIMOSA)** — A non-profit organization directed to facilitating the development of open exchange of equipment condition, maintenance, and lifetime management information.

**Macro** — A set of keystrokes and instructions, in applications, recorded and saved under a short key code or macro name. When the key code is typed, or the macro name is used, the program carries out the instructions of the macro.

**Magnetic Contactor** — A device used to open or close one or more sets of electrical contacts. It is actuated by either energizing or de-energizing an electromagnet within the device.

**Magnetic Filter** — A filter element that, in addition to its filter medium, has a magnet, or magnets, incorporated into its structure to attract and hold ferromagnetic particles.

**Magnetic Plug** — A device strategically located in the flow stream to collect a representative sample of wear debris circulating in the system. For example, engine swarf, bearing flakes, and fatigue chunks. The rate of buildup of wear debris reflects degradation of critical surfaces.

| | |
|---|---|
| **Mainframe Computer** | A high-level, typically large and expensive computer designed to handle intensive computational tasks. Mainframe computers are characterized by their ability to simultaneously support many users connected to the computer by terminals. |
| **Maintainability** | The ease and speed with which a maintenance activity can be carried out on an item. Maintainability is a function of equipment design and maintenance task design. It is usually measured as mean time to repair (MTTR). |
| **Maintainability Engineering** | The set of technical processes that apply maintainability theory to establish system maintainability requirements, allocate these requirements down to system elements, and predict and verify system maintainability performance. |
| **Maintenance** | All actions necessary for retaining an item in, or restoring it to, a specified condition. |
| **Maintenance Action** | One or more tasks necessary to retain an item in, or restore it to, a specified condition. A maintenance action includes corrective, and certain preventive and predictive, maintenance tasks that interrupt the asset function. |
| **Maintenance and Reliability Best Practices** | Maintenance and reliability practices which have been demonstrated by organizations who are leaders in their industry. These organizations are quality producers, with competitive prices in their industry. |
| **Maintenance Audit** | A formal review of maintenance management practices and results carried out by an independent third party for the purposes of evaluating performance, identifying areas of strength, weaknesses, and opportunities for improvement. |
| **Maintenance Backlog** | Maintenance tasks that are essential to repair, or prevent, equipment failures that have not been completed yet. |
| **Maintenance Budget** | The estimated cost of performing maintenance tasks and activities during a specified period. |
| **Maintenance Budget Compliance** | A metric based on the comparison of actual cost of maintenance to the estimated budget. |

| | |
|---|---|
| **Maintenance Cost** | Expenditures for maintenance labor (including maintenance performed by operators, e.g., TPM), materials, contractors, services, and resources. Includes all maintenance expenses for outages/shutdowns/turnarounds as well as normal operating times. Includes capital expenditures directly related to end-of-life machinery replacement. Does not include capital expenditures for plant expansions or improvements. |
| **Maintenance Cost per RAV** | The amount of money spent maintaining assets, divided by the replacement asset value (RAV) of the assets being maintained, expressed as a percentage. Usually measured annually. |
| **Maintenance Craft Worker** | The worker responsible for executing maintenance work orders (e.g. electrician, mechanic, or maintenance technician.). |
| **Maintenance Employees** | All personnel, salaried and hourly, direct and indirect, that are responsible for executing work assignments pertaining to the care (maintenance) of physical assets. Examples of maintenance personnel are superintendents, managers, engineers, general foremen, foremen, supervisors, planners, schedulers, clerical assistants, and craft workers. Synonymous with *maintenance personnel* and *maintenance workforce*. |
| **Maintenance Engineering** | A staff function whose prime responsibility is to ensure that maintenance techniques are effective, that equipment is designed and modified to improve maintainability, that ongoing maintenance technical problems are investigated, and that appropriate corrective and improvement actions are taken. |
| **Maintenance History** | A chronological list of all maintenance performed on an asset. Synonymous with *equipment repair history* and *maintenance record*. |
| **Maintenance Hourly Headcount** | The headcount of hourly maintenance personnel who are employed full time performing maintenance at a site. |

**Maintenance Labor Cost**  Maintenance labor hours multiplied by the labor rate, plus any production incentive (but not profit sharing). Includes maintenance labor costs for normal operating times, outages/shutdowns/turnarounds and capital expenditures directly related to end-of-life machinery replacement. Does not include labor used for capital expenditures for plant expansions or improvements. Typically, it does not include temporary contractor labor cost.

**Maintenance Labor Hours**  The amount of maintenance labor expressed in hours. Includes maintenance labor during normal operating times, outages/shutdowns/turnarounds, and for capital expenditures directly related to end-of-life machinery replacement. Does not include labor hours used for capital expenditures for plant expansions or improvements. Typically, it does not include temporary contractor labor hours.

**Maintenance Management**  All maintenance line supervisors, other than those supervisors that predominantly have crafts reporting to them.

**Maintenance Management Systems**  Automated software systems for handling maintenance work orders, as well as associated inventory, purchasing, accounting, and human-resources functions. In some industries, particularly process production, maintenance management systems can play a leading role as the enterprise system. Also used by manufacturers looking to optimize use of capital-intensive equipment. Some maintenance systems have their own financial and purchasing functions. In other cases, integration with ERP is used to provide these functions.

**Maintenance Materials Cost**  The cost of all materials, spare parts, supplies, etc., consumed for maintaining equipment and facility, including materials purchased for maintenance by contractors and excluding materials for capital projects.

**Maintenance Overhead Cost**  The costs that include craft worker benefits, all pay and benefits for maintenance support staff and support personnel (i.e., clerks, storeroom personnel, etc.), computer costs, and other related maintenance support costs.

**Maintenance Policy**  A statement of principle used to guide maintenance management decision making.

| | |
|---|---|
| **Maintenance Procedures** | Maintenance work instructions with details that describe the work and how it will be carried out, including standardized listing of parts, material, consumable requirements, craft and skill needs, etc. Apply to preventive maintenance or repair. |
| **Maintenance Program Plan** | A management plan that addresses the maintenance needs of the organization. It is focused on asset requirements, production capabilities, and life cycle cost by selecting a strategic framework of planned maintenance actions that result in the desired asset operational performance levels at optimum cost. |
| **Maintenance Route** | An established route through a facility along which a maintainer carries out proactive maintenance, detective maintenance, and minor repairs on a routine basis. |
| **Maintenance Schedule** | A list of planned maintenance tasks, with expected start times and durations of each task, to be performed during a given time period. Schedules can apply to different time periods (daily, weekly, etc.). |
| **Maintenance Shutdown** | A period of time during which a plant, department, process, or asset is removed from service specifically for maintenance. |
| **Maintenance Shutdown Costs** | The total costs incurred to prepare and execute all planned maintenance shutdown or outage activities. Including:<br>• Staff costs incurred for planning and management of the maintenance activities performed during the shutdown.<br>• Costs for temporary facilities and rental equipment directly tied to maintenance activities performed during the shutdown.<br><br>Costs associated with capital project expansions or improvements that are performed during the shutdown are not included. The maintenance Shutdown costs are determined and reported for a specific time period, e.g. monthly, quarterly, annually. |
| **Maintenance Strategy** | A long-term plan covering all aspects of maintenance management which sets the direction and contains firm action plans for achieving a desired future state for the maintenance function. |

| | |
|---|---|
| **Maintenance Task** | The maintenance effort necessary for retaining an item in, or restoring it to, a specified condition. |
| **Maintenance Training Costs** | The expenditures for formal training that maintenance personnel receive during a specified time period. |
| **Maintenance Training Hours** | The number of hours of formal training that maintenance personnel receive during a specified time period. |
| **Maintenance Unit Cost** | The total maintenance cost required for an asset or facility to generate a unit of production. |
| **Maintenance Window** | The time frame in which maintenance work can be performed without incurring any unplanned production losses. |
| **Maintenance Work Order** | A formal document to identify, control, and record maintenance related work as it goes through the organization's work flow process. Usually created in the CMMS. |
| **Maintenance Work Order System** | A means of requesting maintenance service, planning, scheduling, controlling work, and focusing field data to create information. |
| **Maintenance Work Request** | A formal or informal document for requesting maintenance related work. |
| **Maintenance, Repair, and Operating Supplies (MRO)** | Items used in support of general operations and maintenance such as maintenance supplies, spare parts, and consumables used in the manufacturing process and supporting operations. |
| **Major Defect** | A single defect that can cause equipment breakdown and operational losses. |
| **Major Repairs** | Extensive, non-routine, scheduled repairs requiring deliberate shutdown of equipment, the use of a repair crew possibly covering several elapsed shifts, significant materials, rigging, and, if needed, the use of lifting equipment. |

**Malcolm Baldrige National Quality Award (MBNQA)**

An award established by the U.S. Congress in 1987 to raise awareness of quality management and recognize U.S. companies that have implemented successful quality management systems. Two awards may be given annually in each of five categories:

1. Manufacturing
2. Service
3. Small business
4. Education
5. Healthcare

The award is named after the late Secretary of Commerce, Malcolm Baldrige, a proponent of quality management. The U.S. Commerce Department's National Institute of Standards and Technology manages the award, and ASQ administers it.

**Man to Part Storage System**

A type of material storage system where personnel go to the storage system to retrieve parts.

**Management**

The process of achieving organizational goals through engaging in the four major functions of:

1. Planning
2. Organizing
3. Leading
4. Controlling

**Management By Objectives (MBO)**

A process through which specific goals are set collaboratively for the organization as a whole, and every unit and individual within it. The goals are then used as a basis for planning, managing organizational activities, and assessing and rewarding contributions.

**Management By Walking Around (MBWA)**

A practice whereby managers frequently tour areas for which they are responsible, talk to various employees, and encourage upward communication.

**Management Information System (MIS)**

A computer-based information system that produces routine reports and on-line access to current and historical information needed by managers mainly at the middle and first-line levels.

|  |  |
|---|---|
| **Management Of Change (MOC)** | The process of bringing planned change to an organization. MOC usually means leading an organization through a series of steps to meet a defined goal. Synonymous with *change management*. |
| **Management Review** | A periodic meeting of management at which the status and effectiveness of the organization's quality management system is reviewed. |
| **Manager** | An individual charged with the responsibility for managing resources and processes. |
| **Manifold** | A filter assembly, containing multiple ports and integral relating components, which services more than one fluid circuit. |
| **Manifold Filter** | A filter in which the inlet and outlet port axes are at right angles, and the filter element axis is parallel to either port axis. |
| **Manipulated Variable** | The quantity or condition, in process control, that is altered by a controlling action in order to change the value of the regulated condition. |
| **Manufacturing Cost** | Costs, including quality-related, direct and indirect labor, equipment repair and maintenance, and other manufacturing support and overhead, directly associated with manufacturing operations. It does not include purchased-materials costs or costs related to sales and other nonproduction functions. |
| **Manufacturing Cycle Time** | The length of time from the start of production and assembly operations, for a particular product, to the completion of all manufacturing, assembly, and testing for that product or specific customer order. Does not include front-end order-entry time or engineering time spent on customized configuration of nonstandard items. |

**Manufacturing Execution System (MES)** — A software-based system that provides a link between planning and administrative systems and the shop floor. It can link MRP II-generated production schedules to direct process-control software. An element of computer-integrated manufacturing, MES encompasses such functions as planning and scheduling, production tracking and monitoring, equipment control, maintaining product histories (verifying and recording activities at each stage of production), and quality management.

**Manufacturing Resource Planning (MRP II)** — A computer-based information system that integrates the production planning and control activities of basic MRP systems with related financial, accounting, personnel, engineering, and marketing information.

**Manufacturing Variation** — Process variation represented by a normal distribution curve that shows the characteristic variation expected or measured during a manufacturing or assembly operation.

**Man-Way** — An opening in a vessel through which a worker can enter.

**Marginal Cost** — The increase or decrease in the total cost of a production run from making one additional unit of an item. It is computed in situations where breakeven point has been reached.

**Markov Analysis** — A reliability analysis method used to analyze repairable and non-repairable systems with constant failure rates.

**Maslow's Hierarchy of Needs** — A theory of motivation developed by Abraham Maslow in which a person's needs arise in an ordered sequence from the following five categories:

1. Physical needs
2. Safety needs
3. Love needs
4. Esteem needs
5. Self-actualization needs

**Master Black Belt (MBB)** — Six Sigma or quality experts responsible for strategic implementations within the business. The Master Black Belt is qualified to teach other Six Sigma facilitators methodologies, tools and applications in all functions and levels of the company and is a resource for utilizing statistical process control within processes.

**Master Budget**  The total budget package of an organization, including both the operating and financial budgets. Synonymous with *profit plan*.

**Material Review Board**  A quality control team that has the responsibility and authority to deal with materials that do not conform to fitness-for-use specifications.

**Material Safety Data Sheet (MSDS)**  A formal document containing important information about the characteristics and actual or potential hazards of a substance. It identifies:

- Manufacturer – including name, address, phone, and fax Chemical identity
- Hazardous ingredients
- Physical and chemical properties
- Fire and explosion data
- Reactivity data
- Health hazards data
- Exposure limits data
- Precautions for safe storage and handling
- Need for protective gear
- Spill control, cleanup, and disposal procedures

Mandated by the US Occupational Safety and Health Administration (OSHA), it is also used in many other countries in one form or another.

**Materials Resources Planning (MRP)**  Computerized ordering and scheduling system for manufacturing and fabrication industries. It uses bill of materials data, inventory data, and the master production schedule to project what material is required, when, and in what quantity.

**Matrix Diagram**  A planning tool for displaying the relationships among various data sets.

| | |
|---|---|
| **Matrix Organization** | A multifunctional team structure that facilitates horizontal flow of authority, in addition to normal (vertical) flow, by abandoning the "one person, one boss" rule of conventional organizations. Used mainly in management of large projects or product development processes, it draws employees from different functional disciplines (accounting, engineering, marketing, etc.) for assignment to a team without removing them from their respective positions. |
| **Maximum Rated Load** | Total of all loads, including the working load, the weight of the scaffold, and such other loads as may be reasonably anticipated. |
| **Maximum Suction Lift** | The height (approx. 25 ft') that water can be lifted by a centrifugal pump in actual conditions, taking into consideration altitude, friction loss, temperature, suspended particles, and the inability to create a perfect vacuum. |
| **Mean** | Measure of central tendency. The arithmetic average of all measurements in a data set. |
| **Mean Down Time (MDT)** | The average total downtime required to restore an asset to its full operational capabilities. MDT includes the time from the reporting of an asset being down to the time the asset is given back to operations/production to operate. MDT includes administrative time of reporting, logistics, materials procurement and lock-out/tag-out of equipment, etc. for repair or preventive maintenance. |
| **Mean Time Between Events (MTBE)** | The average length of time between one maintenance event (action) and another event for an asset or component. The events can be preventive maintenance work or repair task. Trending the MTBE metric helps to optimize visits to a specific asset or area. |
| **Mean Time Between Failures (MTBF)** | The average length of time between one failure and another for an asset or component. MTBF is usually used for repairable assets of similar type. Another term, Mean Time to Failure (MTTF) is usually used for non-repairable assets, i.e. light bulbs, rocket engines etc. Both terms are used as a measure of asset reliability. These terms are synonymous with *mean life*. MTBF is the reciprocal of the Failure Rate ($\lambda$). |

| | |
|---|---|
| **Mean Time Between Maintenance (MTBM)** | The average length of time between one maintenance action and another for an asset or component. These actions could be either corrective or preventive maintenance. The metric is applied only for maintenance actions which require or result in function interruption. |
| **Mean Time Between Repair (MTBR)** | The calendar time from the restoration of one failure to the onset of the next, excluding the repair time. |
| **Mean Time To Failure (MTTF)** | The average length of operating time to failure of a non-repairable asset or component. |
| **Mean Time To Repair (MTTR)** | The average time needed to restore an asset to its full operational capabilities after a failure. MTTR is a measure of asset maintainability. |
| **Measurand** | A particular quantity that is subject to measurement. |
| **Measure** | The criteria, metric, or means to which a comparison is made with output. |
| **Measurement** | The act or process of quantitatively comparing results with requirements. |
| **Measurement Error** | The difference between the actual value and the measured value of a parameter. |
| **Measurement System** | Measurement process description stating which components are included (apparatus, method, environmental conditions, (training of) operators) thereby identifying which sources of measurement variation are present. |
| **Measurement Systems Analysis** | The study of an organization's measurement system to determine its reliability. An improperly functioning measurement system can introduce variability that negatively affects process capability. |
| **Measurement Uncertainty (U)** | A parameter associated with the result of measurement that characterizes the dispersion of values that could reasonably be attributed to the measurand. |

**Mechanical Integrity (MI)** An element of process safety management (PSM). Mechanical integrity (MI) ensures that equipment does not fail in a way that causes or affects a release of covered chemicals. Equipment means hardware that helps contain the chemicals in the process. MI covers the proper design, fabrication, construction, installation and operation of equipment throughout the entire process life cycle.

**Mechanical Seal** A device to prevent leakage of fluids, at shaft entry points, from inside equipment to the outside.

**Mechatronics** An approach to engineering design and production based on the integration of mechanical and electrical engineering, along with computer science and software engineering.

**Median** The middle number, or center value, of a set of data in which all the data are arranged in sequence.

**Medium** The porous material that performs the actual process of filtration. The plural of this word is media.

**Megger** A commonly used name for a megohmmeter.

**Megohmmeter** A meter used to measure high resistance under a direct current (DC). Synonymous with *megger*.

**Megohmmeter Test** A test used to measure resistance, whereby a constant DC voltage source is applied to the resistance to be measured and the resulting current is read on a highly sensitive ammeter circuit (megohmmeter) that can display the resistance value. Synonymous with *megger test*.

**Memory** A device where information can be stored and retrieved. In the most general sense, memory can refer to external storage such as disk drives or tape drives. In common usage, it refers only to a computer's main memory, the fast semiconductor storage (RAM) directly connected to the processor.

**Menu** A list of options from which a user can make a selection in order to perform a desired action, such as choosing a command or applying a particular format to part of a document.

**Metal Oxides** — Oxidized ferrous particles which are very old or have been recently produced by conditions of inadequate lubrication.

**Metric** — A standard for measurement.

**Metrics** — Numerical measures used to assess performance and effectiveness in a specific area. Metrics are the heart of a good, customer-focused process management system and any program directed at continuous improvement.

**Metrology** — The science of weights and measures or of measurement. A system of weights and measures.

**Microcomputer** — A computer built around a single-chip microprocessor.

**Micron** — A unit of length. One micron = 39 millionths of an inch (.000039"). Contaminant size is usually described in microns. Relatively speaking, a grain of salt is about 60 microns and the eye can see particles to about 40 microns. Many hydraulic filters are required to be efficient in capturing a substantial percentage of contaminant particles as small as 5 microns. Synonymous with *micrometer*, and exhibited as μm.

**Microprocessor** — A central processing unit (CPU) on a single chip.

**Microscope Method** — A method of particle counting which measures or sizes particles using an optical microscope.

**Mil** — Length or displacement equal to 0.001 inch or 0.0254 mm.

**MIL-HDBK** — A U.S. Military Handbook.

**MIL-HDBK-217** — A reliability prediction standard originally developed for defense and aerospace related organizations, but later adopted by many commercial and industrial companies. Often referred to simply as 217, MIL-HDBK-217 includes mathematical reliability models for nearly all types of electrical and electronic components.

**MIL-HDBK-472** — The accepted industry standard for maintainability prediction analysis.

| | |
|---|---|
| **MIL-Q-9858A** | A military standard that describes quality program requirements. |
| **MIL-STD** | A U.S. Military Standard. |
| **MIL-STD-105E** | A military standard that describes the sampling procedures and tables for inspection by attributes. |
| **MIL-STD-1629** | A long recognized standard that describes a method of performing Failure Mode, Effects, and Criticality Analyses. It is used by government, military, and commercial organizations worldwide. MIL-STD-1629 provides calculations for criticality that allows ranking of failure modes dependent on severity classification. |
| **MIL-STD-45662A** | A military standard that describes the requirements for creating and maintaining a calibration system for measurement and test equipment. |
| **Milestone** | A significant point or event in a project. |
| **Milestone Chart** | A scheduling technique used to show the start and completion of milestones on a time-scale chart. Normally, planned events are expressed on a time-scale chart and are shown as solid triangles. Rescheduled or slipped events are usually displayed as hollow diamond symbols. When the late milestones are completed, the diamonds are filled in. |
| **Milk Run** | A method of routing a delivery vehicle that allows it to make pickups or drop-offs at multiple locations on a single travel loop, as opposed to making separate trips to each location. |
| **Mind Map** | A diagram used to represent words, ideas, tasks, or other items linked to and arranged around a central key word or idea. Mind maps are used to generate, visualize, structure, and classify ideas, and as an aid in study, organization, problem solving, decision making, and writing. |
| **Mind Mapping** | A graphical technique for visualizing connections between several ideas or pieces of information using mind maps. |
| **Mineral Oil** | An oil derived from a mineral source, such as petroleum, as opposed to oils derived from plants and animals. |

| | |
|---|---|
| **Minicomputer** | A mid-level computer built to perform complex computations while dealing efficiently with a high level of input and output from users connected via terminals. |
| **Minor Defect** | A single defect that cannot cause losses on its own but may contribute to losses in combination with other minor defects. |
| **Minor Repairs** | Repairs usually performed by one worker using hand tools, a few parts, and usually completed in less than one half shift. |
| **Minor Stoppage Losses** | A minor stoppage loss occurs when a machine is stopped due to a temporary problem, such as a jammed part or feed stoppage (e.g. a clogged inlet chute). The machine resumes production after the problem has been cleared. |
| **Misalignment** | A condition of being badly or improperly aligned. |
| **Miscible** | Capable of being mixed in any concentration without separation of phases, e.g., water and ethyl alcohol are miscible. |
| **Mishap** | An event in which loss is experienced. |
| **Mission** | An organization's purpose. |
| **Mission Profile** | A time-phased description of the events and environments an item experiences from initiation to completion of a specified mission, to include the criteria or mission success or critical failures. |
| **Mission Statement** | A broad declaration of the basic, unique purpose and scope of operations that distinguishes the organization from others of its type. |
| **Mistake Proofing** | The use of any automatic device or method that either makes it impossible for an error to occur, or makes the error immediately obvious once it has occurred. Synonymous with *error proofing* and *poka-yoke (Japanese)*. |
| **MKSA System** | A system of units using the meter, kilogram, second, and ampere as the basic units of length, mass, time, and electric current. |

| | |
|---|---|
| **Modal Analysis** | The process of breaking complex structural motion into individual vibration modes. Resembles frequency domain analysis that breaks complex vibration down to component frequencies. |
| **Modal Probability** | The conditional probability that an item will fail in a particular mode, given that the item has failed. |
| **Mode** | The value occurring most frequently in a data set. |
| **Model** | A representation of a process or system that attempts to relate the most important variables in the system in such a way that analysis of the model leads to insights into the system. |
| **Modeling** | The mathematical characterization of a process, object, or concept to enable the manipulation of variables so as to simulate typical behavior in programmed situations. |
| **Modem** | A communications device that converts between digital data from a computer or terminal and analog audio signals that can pass through a standard telephone line. Short for modulator/demodulator. |
| **Modification** | Any activity carried out on an asset which increases the capability of that asset to perform its required functions. |
| **Modification Time** | The time needed to make modifications or changes to the asset, including capital work. |
| **Modular Drawer Storage** | A lockable storage cabinet containing multiple custom-divided drawers that closely match the specific part or tool configuration requirements. |
| **Module** | A self-contained hardware component that can provide a complete function to a system, and which can be interchanged with other modules that provide similar functions. |
| **Modulus of Elasticity** | The initial slope of the stress vs. strain curve, where Hooke's Law applies, before the elastic limit is reached. |
| **Moly** | Molybdenum disulfide, a solid lubricant and friction reducer, colloidally dispersed in some oils and greases. |

**Monitor** — Observation and recording of a specified parameter that can be used to assess equipment condition. For example, vibration monitoring, thickness measurements, or oil analysis.

**Monitor – Computer** — The device on which images generated by the computer's video adapter are displayed.

**Monte Carlo Method** — A simulation technique by which approximate evaluations are obtained in the solution of mathematical expressions so as to determine the range or optimum value. The technique consists of simulating an experiment to determine some probabilistic property of a system, or population of objects or events, by the use of random sampling applied to the components of the system, objects, or events.

**Motherboard** — The main circuit board containing the primary components of a computer system. This board contains the processor, main memory, support circuitry, and bus controller and connector.

**Motor – Electric** — A device which converts electric power into mechanical force and motion. It usually provides rotary mechanical motion.

**Motor Circuit Analysis (MCA)** — An analysis measuring the phase impedance in the circuitry between the motor and the motor control center. The readings are compared to standards in order to detect imbalances in resistance or impedance. These imbalances may be caused by corrosion, mismatched cables, or current leakage in the circuit. Motor defects such as broken bars, shorts, and eccentricity can also be detected.

**Motor Control Valve** — A valve incorporated in automatic control systems to regulate the rate of flow of material through a section of pipe. It is actuated either mechanically, electrically, or by gas pressure from a control instrument.

**Motor Current Analysis (MCA)** — An analysis that monitors the motor current during start-up (surge-current) and the current trace over time (decay) to detect friction forces.

**Motor Current Signature Analysis (MCSA)** — A method of detecting the presence of broken or cracked rotor bars, cracked end rings, rotor-stator eccentricity, and other defects. Motor current spectrum in both time and frequency domains are collected with a clamp-on ammeter and Fast Fourier Transform (FFT) analyzer. The motor must be energized and loaded during the test.

**Motor Flux Analysis (MFA)** — A method using a magnetic field sensing coil attached to the external surface of operating electrical motors to monitor conditions inside the motor. Sensor signals from "leakage" flux are subjected to Fast Fourier Transformation and presentation for identification of frequency "spikes" indicative of internal defects such as broken rotor bars, and stator phase-to-phase and turn-to-turn faults.

**Motor Normalized Temperature Analysis (MNTA)** — A method using devices, such as low cost infrared radiometers, to measure temperatures at specific points on the outside of operating electric motor casings to give indication of internal defects such as degraded bearings, rotor faults, stator winding faults, clogged air passages, unbalanced motor currents, and couplings in need of realignment or lubrication.

**MRO Store** — A maintenance, repair and operations (MRO) store. It stocks all the material and spare parts required to support maintenance and operations.

**Muda** — A Japanese term for waste. Any activity that consumes resources but creates no value for the customer.

There are seven recognized wastes:

1. Overproduction – production excesses
2. Unnecessary transportation – movement of products
3. Inventory – excessive materials or storage area
4. Motion – walking to get parts, poor ergonomics
5. Defects
6. Over-processing – more work done than the customer requires
7. Waiting

**Multicraft** — A person with multiple skill sets to perform a job effectively.

| | |
|---|---|
| **Multigrade Oil** | An oil that meets the requirements of more than one SAE viscosity grade classification and may therefore be suitable for use over a wider temperature range than a single-grade oil. |
| **Multipass Test** | Filter performance tests in which the contaminated fluid is allowed to re-circulate through the filter for the duration of the test. Contaminant is usually added to the test fluid during the test. The test is used to determine the Beta-Ratio (q.v.) of an element. Synonymous with *recirculation test*. |
| **Multiple Regression** | A statistical procedure to determine a relationship between the mean of a dependent random variable and the given value of one or more independent variables. |
| **MUltipleXer (MUX)** | A device for funneling several different streams of data over a common communications line. Multiplexers are used either to attach many communications lines to a smaller number of communications ports or to attach a large number of communications ports to a smaller number of communications lines. |
| **Multitasking** | The performance of multiple tasks simultaneously. |
| **Multivariate Control Chart** | A control chart for evaluating the stability of a process in terms of the levels of two or more variables or characteristics. |
| **Multi-Voting** | Typically used after brainstorming, multi-voting narrows a large number of possibilities to a smaller list of the top priorities (or to a final selection) ranked in importance by participants. Multi-voting is preferable to straight voting because it allows an item that is favored by all to rise to the top. |
| **Mura** | A Japanese term for unevenness or inconsistencies in physical matters. |
| **Muri** | A Japanese term for unreasonableness of work that an organization imposes on workers. |
| **Myers-Briggs Type Indicator** | A methodology and an instrument for identifying an individual's personality type based on Carl Jung's theory of personality preferences. |

# N

| | |
|---|---|
| **n, N** | Number of observations in a sample or population. |
| **Nameplate** | The plate on the outside of a piece of equipment that describes the rating, capacity, horsepower, manufacturer, serial number, model number, etc. |
| **Nanotechnology** | The engineering of functional systems and devices at the molecular level. |
| **Naphthenic** | A type of petroleum fluid derived from naphthenic crude oil containing a high proportion of closed-ring methylene groups. |
| **Narrowband Trending** | Monitoring of total energy for a specific bandwidth of vibration frequencies. |
| **National Institute of Standards and Technology (NIST)** | An agency of the U.S. Department of Commerce that develops and promotes measurements, standards, and technology. NIST also manages the Malcolm Baldrige National Quality Award. |
| **Natural Frequency** | The frequency of an undamped system's free vibration. The frequency of any of the normal modes of vibration. Natural frequency drops when damping is present. |
| **Natural Team** | A team of individuals drawn from a single work group. Similar to a process improvement team except that it is not cross functional in composition and it is usually permanent. |
| **National Electrical Code (NEC)** | The National Electrical Code (NEC) is a United States standard for the safe installation of electrical wiring and equipment. |
| **Needs Analysis** | An examination of the existing need for training within an organization to identify performance areas or programs within an organization where training should be applied. Synonymous with *needs assessment*. |

| | |
|---|---|
| **Net Positive Suction Head (NPSH)** | The difference between the suction pressure of a pump and the vapor pressure of the fluid, measured at the impeller inlet. |
| **Net Present Value (NPV)** | The current value of future revenue based on the time value of money. |
| **Network** | A group of computers and associated devices that are connected by communications facilities. A network can involve permanent connections, such as cables, or temporary connections made through telephone or other communication links. |
| **Network Diagram** | A schematic display of the logical relationships of project activities, usually drawn from left to right to reflect the project. Synonymous with *logic diagram*. |
| **Networking** | A meeting of independent participants who develop a degree of interdependence and share a coherent set of values and interests. |
| **Neural Network** | A type of artificial-intelligence system modeled after the neurons (nerve cells) in a biological nervous system and intended to simulate the way a brain processes information, learns, and remembers. |
| **Neutral – Acidity** | A measure of pH equal to 7.0. Neither acid nor alkaline. |
| **Neutralization Number** | A measure of the total acidity or basicity of an oil. This includes organic or inorganic acids or bases or a combination thereof (ASTM Designation D974-58T). |
| **Newtonian Fluid** | A fluid with a constant viscosity at a given temperature regardless of the rate of shear. Single-grade oils are Newtonian fluids. Multigrade oils are Non-Newtonian fluids because viscosity varies with shear rate. |
| **Newton's Law of Cooling** | The rate of heat transfer for a cooling object is proportional to the temperature difference between the object and its surroundings. |
| **Nodal Point** | A point of minimum shaft deflection in a specific mode shape. May readily change location along the shaft axis due to changes in residual imbalance or other forcing function, or from a change in restraint, such as increased bearing clearance. |

**Node**  Junction point joined to some or all of the other dependency lines in a network. An intersection of two or more lines or arrows.

**No Demand**  Time that an asset is not scheduled to be in service at the start of the planning period because of lack of demand for the product or service.

**Noise**  Anything that interferes with, slows down, or reduces the clarity or accuracy of a communication. Thus, superfluous data or words in a message are noise because they detract from its meaning or create process variation. Noise can be in communication, electrical, or control systems. In electrical systems, the noise is in the form of random disturbances caused by circuit components, electromagnetic interference, or weather conditions.

**Noise Reduction**  A process to reduce noise levels.

**Nominal Filtration Rating**  An arbitrary micrometer value indicated by a filter manufacturer. Due to lack of reproducibility this rating is deprecated.

**Nominal Group Technique**  A technique, similar to brainstorming, used by teams to generate ideas on a particular subject. Team members are asked to silently come up with as many ideas as possible and write them down. Each member is then asked to share one idea, which is recorded. After all the ideas are recorded, they are discussed and prioritized by the group.

**Nonconformance**  A deficiency in characteristics, documentation, procedures, quality of material, service, or product that is unacceptable or indeterminate.

**Nonconformity**  The non-fulfillment of a specified requirement.

**Nondestructive Testing (NDT)**  A collection of technologies that provide a means of assessing the integrity, properties, and condition of components or material without damaging or altering them. NDT methods include ultrasonic testing, eddy current testing, radiography, thermography, visual inspection, magnetic particle testing, dye penetrant testing, acoustic testing, and others. Synonymous with *non-destructive examination (NDE)*.

**Non-Intrusive Task** — Actions not requiring process or asset interruption, equipment shutdown, tagout, entry or disassembly.

**Nonlinearity** — The deviation from a best fit straight line of true output vs. actual value being measured.

**Non-Newtonian Fluid** — A fluid, such as a grease or a polymer-containing oil (e.g., multi-grade oil), in which shear stress is not proportional to shear rate.

**Non-Operational Consequences** — A category of failure consequences that do not adversely affect safety, the environment, or operations, but only require repair or replacement of any item(s) that may be affected by the failure.

**Nonparametric Tests** — Nonparametric tests are often used in place of their parametric counterparts when certain assumptions about the underlying population are questionable.

**Non-Repairable Item** — An item or piece of equipment that cannot be repaired because of economic reasons, such as a light bulb, circuit board, or rocket.

**Non-Routine Maintenance** — Maintenance (usually repairs) performed at irregular intervals, with each job unique, and based on inspection, failure, or condition.

**Non-Stock Item** — An inventoried item that is not physically in the storeroom but is identified in the inventory database. It may be stocked in a vendor's warehouse or can be delivered in specified time from the supplier.

**Non-Value Added (NVA)** — Tasks or activities that can be eliminated with no deterioration in product or service functionality, performance, or quality in the eyes of customer. These tasks are waste in the process and don't add any value.

**Nonwoven Medium** — A filter medium composed of a mat of fibers.

**Norm** — The expectation of how a person(s) or process will behave in a given situation based on established rules of conduct or accepted conditions.

| | |
|---|---|
| **Normal Distribution** | The charting of a data set in which most of the data points are concentrated around the average (mean), thus forming a bell shaped curve. Synonymous with *Gaussian distribution*. |
| **Normality Test** | A statistical process used to determine if a sample, or any group of data, fits a standard normal distribution. A normality test can be performed mathematically or graphically. |
| **Norming** | The third stage of the teaming process. The four stages are:<br>1. Forming<br>2. Storming<br>3. Norming<br>4. Performing |
| **North American Industry Classification System (NAICS)** | A coding system of the US, Mexican, and Canadian governments that identifies specific economic sectors. It replaced the US Standard Industrial Classification (SIC) system. |
| **Not Operating** | The state wherein an item is able to function but is not required to function. |
| **Not Invented Here (NIH) Syndrome** | An attitude that prevents individuals and groups from using and benefiting from the ideas of other individuals or groups because of personal pride, cultural, ethnic or national prejudice, or other bias. |
| **np Chart** | A chart used to monitor the number of items defective for a fixed sample size. |
| **Null Hypothesis** | A hypothesis that is formulated without considering any new information, or is formulated based on the belief that what already exists is true. |
| **Numerical Analysis** | The study of algorithms for solving the problems of continuous mathematics. |
| **Nyquist Plot** | A plot of the real part versus the imaginary part of the frequency response function. |

# O

**O & M Manuals**   Operation and maintenance (O&M) manuals provided by the equipment manufacturers or suppliers, or developed by engineering, giving basic details on operation and maintenance of the equipment. They usually include suggested preventive maintenance tasks, troubleshooting guides, and identification of parts and special tools.

**Objective**   A specific statement of a desired short term condition or achievement. Includes measurable end results to be accomplished by specific teams or individuals within time limits.

**Object-Oriented Database**   A flexible database that supports the use of abstract data types, objects, and classes, and that can store a wide range of data, often including sound, video, and graphics, in addition to text and numbers.

**Obliteration**   A synergistic phenomenon of both particle silting and polar adhesion. When water and silt particles co-exist in a fluid containing long-chain molecules, the tendency for valves to undergo obliteration increases.

**Observation**   Data or information gathered by noting a fact, or viewing an occurrence of events (e.g. observations recorded during a work study project).

**Obsolete Stores Items**   Store items or Stock Keeping Units (SKU) that have had no activity for several years and are not designated as safety or critical spares or items.

**Occupational Safety and Health Administration (OSHA)**   A federal law that applies to all employees in the United States who are engaged in interstate commerce. Its purpose is to ensure safe and healthful working conditions by authorizing enforcement of the standards provided under the act.

**Octave**   The interval between two frequencies with a ratio of 2 to 1.

**Off-Line**  An asset or system that has been currently bypassed or disconnected from operation for maintenance and/or repair. Any element of a process that stands independent of its normal flow.

**Off-the-Shelf Item**  Commercially produced, ready-made, standardized, and regularly available equipment, goods, parts, and software.

**Oil**  A greasy, unctuous liquid of vegetable, animal, mineral or synthetic origin.

**Oil Analysis**  The sampling and laboratory analysis of a lubricant's properties, suspended contaminants, and wear debris. Synonymous with *lube oil analysis*.

**Oil Foaming**  The appearance of closely packed bubbles of air and water entrapped in oil which gives the oil a milky appearance. This can deteriorate lubricating properties. Defoaming or antifoaming agents are used to alleviate the problem.

**Oil Ring**  A loose ring, the inner surface of which rides a shaft or journal and dips into a reservoir of lubricant from which it carries the lubricant to the top of a bearing by its rotation with the shaft.

**Oil Separation Test**  A test used to determine the tendency of lubricating grease to separate oil. Applicable standards are ASTM D6184 and ASTM D1742.

**Oil Whirl/Oil Whip**  An unstable, free vibration whereby a fluid-film bearing has insufficient unit loading. Under this condition, the shaft centerline dynamic motion is usually circular in the direction of rotation. Oil whirl occurs at the oil flow velocity within the bearing, usually 40 to 49% of shaft speed. Oil whip occurs when the whirl frequency coincides with a shaft resonant frequency.

**Oiliness**  The property of a lubricant that produces low friction under conditions of boundary lubrication. The lower the friction, the greater the oiliness.

**On-Condition Task**  A maintenance task performed based on the condition of the item or asset as determined by the scheduled inspection. Synonymous with *condition-directed task*.

| | |
|---|---|
| **Online Order Entry System** | A computer-based system that enables distributors, field sales representatives, and customers to place orders directly, over the Internet or a corporate intranet, without intervention by an inside salesperson. |
| **On-the-Job Training (OJT)** | The training an employee receives while actually doing the job, either alone or under the eye of an experienced co-worker. |
| **On-Time Delivery Rate** | The percentage of time that products ordered by customers are received by the specified time or date. |
| **Opaque** | The material property of being impervious to radiant energy. In thermography, an opaque material is one that does not transmit thermal infrared energy. |
| **OPC Foundation** | An organization dedicated to ensuring interoperability in automation by creating and maintaining open specifications that standardize the communication of acquired process data, alarm and event records, historical data, and batch data to multi-vendor enterprise systems and between production devices. Production devices include sensors, instruments, PLCs, RTUs, DCSs, HMIs, historians, trending subsystems, alarm subsystems, etc., as used in the process industry, manufacturing, and in acquiring and transporting oil, gas, and minerals. |
| **OPC Standards** | Standards promulgated by the OPC Foundation. |
| **Open Bearing** | A ball bearing that does not have a shield, seal, or guard on either of the two sides of the bearing casing. |
| **Open Bubble Point – Boil Point** | The differential gas pressure at which gas bubbles are profusely emitted from the entire surface of a wetted filter element under specified test conditions. |
| **Open Circuit** | A break in an electrical circuit that prevents normal current flow. |
| **Open Systems** | An approach to computing that stresses the interconnectability of systems based on compliance to established standards. |

**Open-Door Policy** — A management approach that encourages employees to speak freely and regularly to management regarding any aspect of the business or project. This tends to minimize personnel problems and employee dissatisfaction.

**Open-Ended Problem** — A problem without a single correct answer and boundaries that can be challenged.

**Operability** — The ability of an equipment to be operated in its intended manner.

**Operable** — The ability of an asset to perform its intended function.

**Operating** — The state of an asset or device when it is electrically or mechanically activated as indicated in its design and any level of stress. If any portion of the asset is operating, then the entire asset/system may be considered to be operating.

**Operating Budget** — A type of budget involving a statement that presents the financial plan for each responsibility center during the budget period and reflects operating activities involving revenues and expenses.

**Operating Characteristic Curve** — A graph to determine the probability of accepting lots as a function of the lots' or processes' quality level when using various sampling plans. There are three types:

1. Type A curves – which give the probability of acceptance for an individual lot coming from finite production (will not continue in the future).
2. Type B curves – which give the probability of acceptance for lots coming from a continuous process.
3. Type C curves – which (for a continuous sampling plan) give the long run percentage of product accepted during the sampling phase.

**Operating Context** — The circumstances in which a physical asset or system is expected to operate.

**Operating Environment** — The aggregate of all external and internal conditions (such as temperature, humidity, radiation, magnetic and electric fields, shock vibration), natural, man-made, or self-induced, that influence the form, operational performance, reliability or survival of an item.

| | |
|---|---|
| **Operating Hours** | The length of time that an item of equipment is actually operating. |
| **Operating Procedure** | A detailed step-by-step written document, or checklist, used to start up, run, or shutdown an asset in a safe, economical, productive and effective way. |
| **Operating Speed** | The speed at which a piece of equipment or system operates. |
| **Operating System** | The software that controls the allocation and usage of hardware resources such as memory, central processing unit (CPU) time, disk space, and peripheral devices. The operating system is the foundation software on which applications depend. |
| **Operating Time** | A particular interval of time during which the item or asset is performing its required function. |
| **Operational Consequences** | A category of failure consequences that adversely affect the operational capability of a physical asset or system (output, product quality, customer service, military capability, or operating costs in addition to the cost of repair). |
| **Operational Control System** | A system that runs a company's day-to-day operations. |
| **Operational Limit** | The extremes beyond which a product is not expected to operate. |
| **Operational Readiness** | The ability of an operations unit to respond to its operation plan(s). A function of equipment availability and status, training, etc. |
| **Operations** | An organizational function performing the ongoing execution of activities that produce the same product or provide a repetitive service. For example, production operations and accounting operations. |
| **Operator Based Maintenance (OBM)** | Minor maintenance tasks performed by operators. |

| | |
|---|---|
| **Operator Driven Reliability (ODR)** | A system of involving the equipment operators in improving reliability by having them identify potential equipment problems and failures early. The operator fixes them if they are minor and issues a repair request if they are major. Synonymous with *operator-based reliability* and *operator-based maintenance*. |
| **Operators** | The plant personnel responsible for operating the plant equipment and systems. |
| **Opportunity** | Any area of a product, process, or service that must be right to achieve customer satisfaction. |
| **Opportunity Cost** | The loss of revenue or cost associated with a lost opportunity to invest in a desired manner or to earn income. |
| **Opportunity Maintenance** | Maintenance work that is performed in an unanticipated maintenance window or to take advantage of a planned maintenance window to get more work accomplished than scheduled. |
| **Optimization** | The process of achieving the best possible solution to a problem in terms of a specified objective function. |
| **OR Gate** | A logic gate in which an output occurs if one or more inputs exist. Any single input is necessary and sufficient to cause the output to occur. |
| **Orbit** | The path of a shaft centerline during rotation. The orbit is usually observed on an oscilloscope connected to x- and y-axis displacement sensors. Synonymous with *Lissajous pattern*. |
| **Ordinate** | The vertical (Y) axis of a chart or graph. The horizontal (X) axis is called the abscissa. |
| **Organization** | A group of persons organized for some purpose or to perform some type of work within an enterprise. |
| **Organization Chart** | A diagram or graphic representation of an organization which shows, to varying degrees, functions, responsibilities, people, authority and the relationships among them. |
| **Organizational Development** | The planned, organization-wide process of change designed to improve organizational effectiveness and adaptation (to changing environmental demands). |

| | |
|---|---|
| **Organizational Interfaces** | Formal and informal reporting and working relationships among different organizational units. |
| **Orifice** | A device for restricting the flow through a pipe. The difference in pressure on the two sides of an orifice plate can be used to measure the volume of flow through pipe. |
| **Orifice Meter** | An instrument that measures the flow through a pipe by determining the difference in pressure on the upstream and downstream sides of an orifice plate. |
| **Original Equipment Manufacturer (OEM)** | An equipment manufacturer who sells a product or system built from its own, or other companies', components. The OEM supplies all technical documentation as references to operate and maintain the product or system. |
| **Originator** | A person that writes a work order or other request for maintenance work. |
| **Oscillation** | Variations with time of a quantity such as force, stress, pressure, displacement, velocity, acceleration, or jerk. Usually implies some regularity (as in sinusoidal or complex vibration). |
| **OSHA 1910** | A set of federal regulations containing the occupational safety and health standards for industries. These have been found to be national consensus standards or established federal standards. |
| **OSHA Lost Work Days Injury Case Rate** | An incidence rate of lost work days due to injury which is computed from the formula: *Lost Days Incidence Rate = (number of lost work days caused by injuries and illnesses cases x 200,000)/employee hours worked* |
| **OSHA Recordable Injuries** | Injuries that must be recorded and reported to OSHA but that do not lead to lost work days. |
| **OSHA Recordable Injury Incident Rate** | An incidence rate of injuries and illnesses established by OSHA. It is computed from the formula: *Injury Incidence Rate = (total number of injuries and illnesses cases x 200,00/employee hours worked* Incidence rates can be used to show the relative level of injuries and illnesses among different industries, firms, or operations within a single organization. |

**Out of Spec**   A term that indicates a unit does not meet a given requirement (or specification). Short for *out of specification*.

**Outage**   A term used in some industries (electrical power distribution) to denote when an item or system is not in use.

**Outlier**   A data point that is more than one or two standard deviations from the mean of a data set.

**Out-of-Control Process**   A process in which the statistical measure being evaluated is not in a state of statistical control. In other words, the variations among the observed sampling results can be attributed to a constant system of chance causes.

**Output**   A product, result, or service generated by a process. May be an Input to a successor process.

**Outsourcing**   The process of awarding a contract or otherwise entering into an agreement with a third party, usually a supplier, to perform work that is currently being performed by an organization's employees. Organizations outsource elements of their operations if the service can be performed cheaper, faster, or more effectively by a third party, or if the service is not considered part of their core business (e.g., janitorial services).

**Overall Equipment Effectiveness (OEE)**   A measure of equipment or process performance based on actual availability, performance efficiency, and quality of product or output. OEE is generally expressed as a percentage.

*OEE = availability x performance x quality*

**Overhaul**   A comprehensive examination and restoration of an asset to an acceptable condition.

**Overload**   A load greater than that which a device is able to handle.

**Overtime**   Any hours beyond the normal standard work time.

**Oxidation**   The process where a material chemically combines with oxygen, usually resulting in the formation of rust.

**Oxidation – Tribology**  A chemical process that occurs when oxygen attacks petroleum fluids. The process is accelerated by heat, light, metal catalysts and the presence of water, acids, or solid contaminants. It leads to increased viscosity and deposit formation.

**Oxidation Inhibitor**  A substance added in small quantities to a petroleum product to increase its oxidation resistance, thereby lengthening its service or storage life. An oxidation inhibitor may work in one of these ways:

- By combining with and modifying peroxides (initial oxidation products) to render them harmless.
- By decomposing peroxides.
- By rendering an oxidation catalyst inert.

Synonymous with *anti-oxidant*.

**Oxidation Stability**  The ability of a lubricant to resist natural degradation upon contact with oxygen.

# P

**P&ID** — Piping And Instrumentation Diagram.

**Pallet Rack Storage** — A type of shelving for big and heavy parts in which the parts are stored on pallets.

**Paper Chromatography** — A method which involves placing a drop of fluid on a permeable piece of paper and noting the development and nature of the halos, or rings, surrounding the drop through time.

**Paradigm** — A set of assumptions, concepts, values, and practices that constitutes a way of viewing reality for the community that shares them.

**Paradigm Shift** — A change from one way of thinking to another.

**Paradox** — A statement or group of statements that leads to a contradiction or a situation which defies intuition. An apparent contradiction that actually expresses a non-dual truth. The word paradox is often used interchangeably with contradiction.

**Paraffinic** — A type of petroleum fluid derived from paraffinic crude oil, containing a high proportion of straight chain saturated hydrocarbons. Often susceptible to cold flow problems.

**Parallel Tasks** — Independent tasks that proceed concurrently.

**Parameter** — A constant or coefficient that describes some characteristic of a population.

**Pareto Chart** — A graphical tool for ranking causes from most significant to least significant. Based on the Pareto principle.

| | |
|---|---|
| **Pareto Principle** | A principle that the critical few, about 20% of items such as assets, failures, parts, etc., should receive attention before the insignificant many, usually about 80%. This principle is named after 19th century economist Vilfredo Pareto, who suggested that most effects come from relatively few causes. That is, 80% of the effects come from 20% of the possible causes. Synonymous with *80/20 rule*. |
| **Part to Man Storage System** | A type of parts storage system that retrieves parts automatically and delivers them to the required location. |
| **Partial Discharge** | A type of localized discharge resulting from transient gaseous ionization in an insulation system when the voltage stress exceeds a critical value. Synonymous with *corona discharge*. |
| **Particle Count** | The numbers of particles present greater than a particular micron size per unit volume of fluid. Often stated as particles greater than 10 microns/milliliter. |
| **Particle Density** | An important parameter in establishing an entrained particle's potential to impinge on control surfaces and cause erosion. |
| **Particle Erosion** | Erosion that occurs when fluid-entrained particles moving at high velocity pass through orifices, or impinge on metering surfaces or sharp angle turns. |
| **Particle Impingement Erosion** | A particulate wear process where high velocity, fluid-entrained particles are directed at target surfaces. |
| **Particle Size Distribution** | The quantification of particles by size range. |
| **Particulate** | A state of matter in which solid or liquid substances exist in the form of aggregated molecules or particles. |
| **Partnership** | Both a strategy and a formal relationship between a supplier and a customer that engenders cooperation for the benefit of both parties. |
| **Parts** | All of the supplies, machine parts, and materials needed to repair an asset, or a system in or around an asset. |

| | |
|---|---|
| **Parts Per Million (ppm)** | Concentration by volume of one part of a gas or vapor, or by weight, of a liquid or solid, per million parts. In Six Sigma, the number of times an occurrence (i.e. defect) happens in one million chances. |
| **PAS 55** | The British Standards Institution's Publicly Available Specification for the optimized management of physical assets. |
| **Patch Test** | A method by which a specified volume of fluid is filtered through a membrane filter of known pore structure. All particulate matter in excess of an "average size," determined by the membrane characteristics, is retained on its surface. Thus, the membrane is discolored by an amount proportional to the particulate level of the fluid sample. Visually comparing the test filter with standard patches of known contamination levels determines acceptability for a given fluid. |
| **Path Set** | A group of fault tree initiators which, if none of them occurs, will guarantee that the top events cannot occur. |
| **Payback Method** | A method of evaluating an investment opportunity that provides a measure of the time required to recover the initial amount invested in the project. It focuses on the payback period that is defined as the amount of time a company expects to take before it recovers its initial investment. |
| **Payback Period** | The number of time periods (usually years) it will take the results of a capital investment project to recover the investment from net cash flows. |
| **Pay-for-Performance** | Compensation scheme for employees and team members in which their level of pay is directly tied to specific business goals and management objectives. Organizations adopt this approach to improve individual accountability, align shareholder, management, and employee interests, and enhance performance. |
| **Peak** | The extreme value of a varying quantity, measured from the zero or mean value. A maximum spectral value. |
| **Peak Amplitude** | The value of the highest amplitude measured from a defined reference point. |

| | |
|---|---|
| **Peak-to-Peak Amplitude** | The value of the amplitude as measured from the highest amplitude peak to the lowest amplitude peak. |
| **Pending Work** | Work that has been issued for execution but is not yet completed. Maintenance work in process. |
| **Percent Chart** | A control chart for evaluating the stability of a process in terms of the percentage of the total number of units in a sample in which an event of a given classification occurs. Synonymous with *proportion chart*. |
| **Percent Planned Work** | The percentage of total work (in labor hours or number of work orders) performed in a given time period which has been planned in advance. |
| **Performance – Equipment** | How equipment is performing compared to its designed capability and/or reliability objectives. |
| **Performance – Personnel** | The result of activities of a person, team, organization, or equipment over a given period of time as compared to a standard or specified goals. |
| **Performance Efficiency** | Actual output of a person, equipment, or system compared with the desired or planned output, usually expressed as a percentage. |
| **Performance Indicator** | A variable, derived from one or more measurable parameters, which, when compared with a target level or trend, provides an indication of the degree of control being exercised over a process (e.g. work efficiency, equipment availability). Synonymous with *performance indices*. |
| **Performance Measurement** | A process of assessing progress toward achieving predetermined goals, including information on the efficiency with which resources are transformed into goods and services, outputs, the quality of those outputs, how well they are delivered to the customer and the extent to which customers are satisfied. |
| **Performance Standard** | The metric against which a complete action is compared. |

| | |
|---|---|
| **Performance Trending** | A process of assessing progress toward achieving predetermined goals, including information on the efficiency with which resources are transformed into goods and services, outputs, the quality of those outputs, how well they are delivered to the customer and the extent to which customers are satisfied. |
| **Performance-Based Requirements** | Requirements that describe what the product should do, how it should perform, the environment in which it should operate, and interface and interchangeability characteristics. |
| **Performing** | The fourth stage of the teaming process. |
| **Period** | A time-frame during which performance of assets or an organization is measured. A period can be expressed in time units (hours, months, etc.) or in cycles (revolutions, radians, number of cycles). |
| **Period – Vibration** | The interval of time over which a cyclic vibration repeats itself. |
| **Periodic Maintenance** | Cyclic maintenance actions carried out at regular intervals, based on repair history data, use, or elapsed time. |
| **Periodic Vibration** | An oscillation whose waveform regularly repeats. Compare with probabilistic vibration. |
| **Permeability** | The relationship of flow per unit area to differential pressure across a filter medium. |
| **Personal Digital Assistant (PDA)** | A lightweight palmtop computer designed to provide specific functions like personal organization (calendar, note taking, database, calculator, and so on) as well as communication. |
| **Personal Protective Equipment (PPE)** | Protective clothing, helmets, goggles, or other garments designed to protect the wearer's body or clothing from injury, from electrical hazards, heat, chemicals, and infection, for job-related occupational safety and health purposes. |

| | |
|---|---|
| **PERT Chart** | A graphical project management technique that shows the time taken by each component of a project, and the total time required for its completion. The program evaluation and review technique (PERT) breaks down the project into events and activities, and lays down their proper sequence, relationships, and duration in the form of a network. Lines connecting the events are called paths, and the longest path resulting from connecting all events is called the critical path. The length (duration) of the critical path is the duration of the project, and any delay occurring along it delays the whole project. |
| **P-f Interval** | The interval between the point at which a potential failure becomes detectable and the point at which it degrades into a functional failure. Synonymous with *lead time to failure*. |
| **pH** | A measure of alkalinity or acidity in water and water-containing fluids. pH can be used to determine the corrosion-inhibiting characteristic in water-based fluids. Typically, pH > 8.0 is required to inhibit corrosion of iron and ferrous alloys in water-based fluids. |
| **Phase – Chemical** | The state of a material, either liquid, solid, or gas. |
| **Phase – Motor** | The space relationships of windings and changing values of the recurring cycles of AC voltages and currents. |
| **Phase – Vibration** | A measurement of the timing relationship between two signals, or between a specific vibration event and a trigger signal with the same frequency. Phase is often measured as a function of frequency. |
| **Phase Change** | The process matter goes through when it changes from one state to another, solid to liquid, liquid to gas. |
| **Phase Reference Probe** | A device for giving a once-per-shaft-revolution signal. |
| **Phase-to-Ground Insulation** | The insulation between windings as whole and the "ground" or metal part of an electric motor. |
| **Phase-to-Phase Insulation** | The insulation between adjacent coils in different phase groups on an electric motor. |

| | |
|---|---|
| **Physical Configuration Audit (PCA)** | A review of system hardware, drawings, and documentation to ensure the as-built system and as-coded items are accurately reflected in their documentation Usually this audit is performed during commissioning of new or modified assets or for major repairs. |
| **Physical Inventory** | A "hands-on" count of a specifically designated group of items to ensure its accuracy in the system. |
| **Pick List** | A selection of required stores items for a work order or task. Normally used by stores to prepackage the needed materials for use. |
| **Pictograph** | An ideogram that conveys its meaning through its pictorial resemblance to a physical object. Used in statistics to represent statistical data using symbolic figures to match the frequencies of different kinds of data. |
| **PID** | Proportional plus integral plus derivative (control). |
| **Pie Chart** | A circular, radially divided, graphical display indicating proportional data. |
| **Piece Work** | An arrangement where technicians receive pay based on how many units are produced. |
| **PiezoElectric (PE)** | Any material which provides a conversion between mechanical and electrical energy. For a piezoelectric crystal, if mechanical stresses are applied on two opposite faces, electrical charges appear on some other pair of faces. |
| **PiezoElectric (PE) Transducer** | A transducer that produces an electric charge in direct proportion to the vibratory acceleration. |
| **PiezoResistive (PR) Transducer** | A transducer whose electrical output depends upon deformation of its semiconductor resistive element. |
| **Pigeon-Hole** | A small compartment or recess, as in a desk, for holding papers. A cubbyhole. |
| **Pilot Project** | An effort undertaken to test the feasibility of applying results of an initiative on a broader scale. |
| **Pinion Gear** | The smaller of two mating or meshing gears. Can be either the driving or the driven gear. |

| | |
|---|---|
| **Piping and Instrumentation Diagram (P&ID)** | Diagram of the unit, system, or equipment showing control systems such as valves and steam traps, and instruments such as pressure indicators, temperature indicators, and level indicators or controllers. |
| **Pitch** | Rotation in the plane of forward motion about the left-right axis. |
| **Pitting** | A form of extremely localized attack characterized by holes in the metal. Pitting is one of the most destructive and insidious forms of corrosion. Depending on the environment and the material, a pit may take months, or even years, to become visible. |
| **Pixel** | Short for picture (pix) element. One spot in a rectilinear grid of thousands of such spots that are individually "painted" to form an image produced on the screen by a computer, or on paper by a printer. A pixel is the smallest element that display or print hardware and software can manipulate in creating letters, numbers, or graphics. |
| **Plan** | The comprehensive description of maintenance work to be done, including task list, parts, materials and tools required, safety precautions to be observed, permits and other documentation requirements, and estimate of the duration of the work, effort, and costs. |
| **Planck's Curve** | A set of curves that describe the relationship between the temperature of a blackbody and the amount of energy it radiates, as well as the distribution of the wavelength of that energy. |
| **Plan-Do-Check-Act (PDCA) Cycle** | A four-step process for quality improvement. In the first step (plan), a plan to effect improvement is developed. In the second step (do), the plan is carried out, preferably on a small scale. In the third step (check), the effects of the plan are observed. In the last step (act), the results are studied to determine what was learned and what can be predicted. Synonymous with *Shewhart cycle*. |
| **Planned and Scheduled Man Hours** | A predetermined number of man-hours by maintenance staff to be used on planned and scheduled tasks. |
| **Planned Cost** | The estimate of what it should cost to complete a task, project or activity. It includes service and material requirements. Contingency allowances are not included. |

| | |
|---|---|
| **Planned Downtime** | The amount of downtime officially scheduled in the production plan that includes downtime for scheduled maintenance and management activities (e.g., training, meetings). |
| **Planned Maintenance** | Any maintenance activity for which a pre-determined job procedure has been documented, for which all labor, materials, tools, and equipment required to carry out the task have been estimated, and their availability assured before commencement of the task. |
| **Planned Repair Schedule Compliance** | The number of planned repair work orders ( or man-hours) completed from the daily/weekly schedule divided by the total number of work orders (total man-hours) on the schedule. |
| **Planned Work** | Work that has gone through a formal planning process to identify labor, materials, tools, and safety requirements. This information is assembled into a job plan package and communicated to craft workers prior to the start of the work. |
| **Planned Work Executed** | Work that was formally planned and completed. |
| **Planned Work Order Hours** | The planner's estimate of hours needed to complete a work order. |
| **Planner** | A dedicated role with the single function of planning work tasks and activities. |
| **Planner Library** | An efficiency improvement planning tool consisting of hardcopy and/or electronic (including CMMS) storage of job plan information to be consulted for future plans. |
| **Planner-to-Craft Ratio** | The number of maintenance workers a single planner is preparing planned maintenance activities for. |
| **Planning** | Identifying the safety precautions, procedure, tools, skills and time necessary to perform a given task. |
| **Planning Period** | The period of time that defines how the potential capacity will be used. |
| **Planning Variance Index** | The percent of work orders closed where the actual cost varied by within a certain percentage, e.g., +/- 20% from the planned cost. |

**Plant Controllable Costs**   The annual cost of goods sold less raw materials, depreciation, and taxes.

**Plant Engineering**   A staff function whose prime responsibility is to ensure that maintenance techniques used are effective, that equipment is designed and modified to improve maintainability, that ongoing maintenance technical problems are investigated, and that appropriate corrective and improvement actions are taken.

**Plant Replacement Value (PRV)**   The amount of capital that would be required to replace the plant. This is not the book value nor the current cost accounting value nor the costs to build a state of the art replacement. PRV is an estimate of the current costs to replace in kind what now exists, normally the insurance value. A related term is replacement asset value (RAV).

**Plant Workforce**   All personnel (management, supervision, staff, and Line) who are accountable for the successful running of a plant. Exclude corporate level personnel if they are co-located.

**Pleated Filter**   A filter element whose medium consists of a series of uniform folds and has the geometric form of a cylinder, cone, disc, plate, etc. Synonymous with *convoluted filter* and *corrugated filter*.

**Plenum**   An air compartment or chamber to which one or more air ducts are connected. Part of an air handling system that supplies conditioned air, circulates air, or exhausts air.

**PM & PdM Compliance**   The number of preventive maintenance (PM) and predictive maintenance (PdM) work orders completed from the daily or weekly schedule divided by the total number of PM work orders scheduled.

**PM & PdM Corrective Work Orders**   Corrective work orders that are generated from preventive maintenance (PM) or predictive maintenance (PdM).

**PM & PdM Effectiveness**   A measure of the effectiveness of the corrective work that results directly from preventive maintenance (PM) and predictive maintenance (PdM) strategies. The measure is the amount of corrective work identified from PM/PdM work orders that was truly necessary.

| | |
|---|---|
| **PM & PdM Frequency** | Cyclical period of time upon which preventive maintenance (PM) and predictive maintenance (PdM) activities are repeated. |
| **PM & PdM Work Order Backlog** | A measure of all active preventive maintenance (PM) and predictive maintenance (PdM) work orders (i.e., ongoing – not closed) in the system. |
| **PM & PdM Work Orders Overdue** | The number of preventive maintenance (PM) and predictive maintenance (PdM) work orders that are past their required-by date (i.e., overdue). |
| **PM & PdM Yield** | A measure of corrective work that results directly from preventive maintenance (PM) and predictive maintenance (PdM) tasks in place. The measure is the amount of repair and replacement work that is identified when performing PM or PdM work compared to the amount of PM or PdM work being done. |
| **PM Frequency** | The frequency for performing preventive maintenance (PM) work. |
| **PM Optimization** | A process to optimize preventive maintenance (PM) tasks and frequencies to reduce likely failure modes by utilizing tools/techniques such as FMEA, RCM, and CBM. |
| **Pneumatics** | The engineering science pertaining to gaseous pressure and flow. |
| **Point Estimate** | The single value used to estimate a population parameter. Point estimates are commonly referred to as the points at which the interval estimates are centered. These estimates give information about how much associated uncertainty there is. |
| **Point of Diminishing Returns** | The point where an additional action or task provides a benefit, but the benefit would not outweigh the added cost. |
| **Point of Use** | A technique that ensures people have exactly what they need to do their job - the right work instructions, parts, tools and equipment - where and when they need them. |
| **Points of Control** | Areas where process measurement indicates the level of process performance. |

**Poise**	A measure of viscosity numerically equal to the force required to move a plane surface of one square centimeter per second when the surfaces are separated by a layer of fluid one centimeter in thickness. It is the ratio of the shearing stress to the shear rate of a fluid and is expressed in dyne seconds per square centimeter. 1.0 centipoise equals 0.01 poise.

**Poisson Distribution**	A discrete probability distribution that expresses the probability of a number of events occurring in a fixed time period if these events occur with a known average rate and are independent of the time since the last event.

**Poka-Yoke**	A Japanese term that means mistake-proofing. A poka-yoke device is one that prevents incorrect parts from being made or assembled, or easily identifies a flaw or error.

**Polar Compound**	A chemical compound whose molecules exhibit electrically positive characteristics at one extremity and negative characteristics at the other. Polar compounds are used as additives in many petroleum products. Polarity gives certain molecules a strong affinity for solid surfaces. As lubricant additives (oiliness agents), such molecules plate out to form a tenacious, friction-reducing film. Some polar molecules are oil-soluble at one end and water-soluble at the other end. In lubricants, they act as emulsifiers, helping to form stable oil-water emulsions. Such lubricants are said to have good metal-wetting properties. Polar compounds with a strong attraction for solid contaminants act as detergents in engine oils by keeping contaminants finely dispersed.

**Polar Plot**	A vector plot of transient data, amplitude and phase, at varied speed.

**Polarization Index**	A ratio of 10 minute to 1 minute resistance-to-ground (megohmmeter) test values as determined during a polarization test.

**Polarization Index Profile**	The graphical presentation of a series of resistance-to-ground readings taken during a polarization test.

**Polarization Test**	A test used to detect deterioration of insulation, or contaminants, in electric motor or small generator circuits.

| | |
|---|---|
| **Poles** | The number of magnetic poles, in an AC motor, in the stator winding. The number of poles determines the motor's speed. In a DC motor, refers to the number of magnetic poles in the motor. They create the magnetic field in which the armature operates. |
| **Policy** | An overarching plan (direction) for achieving an organization's goals. |
| **Polishing – Bore** | An excessive smoothing of the surface finish of the cylinder bore, or cylinder liner, to a mirror-like appearance in an engine, resulting in depreciation of ring sealing and oil consumption performance. |
| **Polymerization** | The chemical combination of similar type molecules to form larger molecules. |
| **Population** | The complete set of items from which a sample can be drawn. The entire collection of items about which a researcher seeks to draw conclusions. All items in a population share at least one measurable feature. |
| **Pore** | A small channel or opening in a filter medium which allows passage of fluid. |
| **Pore Size Distribution** | The ratio of the number of effective holes of a given size to the total number of effective holes per unit area expressed as a percent and as a function of hole size. |
| **Porosity** | The ratio of pore volume to total volume of a solid (e.g., filter medium) expressed as a percent. |
| **Potential Failure** | An identifiable condition that indicates that a functional failure is either about to occur or is in the process of occurring. |
| **Potentiometer** | An instrument used to measure, or compare, electromotive forces. A variable resistor. |
| **Pour Point** | The lowest temperature at which an oil or distillate fuel is observed to flow, when cooled under conditions prescribed by test method ASTM D 97. The pour point is 3°C (5°F) above the temperature at which the oil in a test vessel shows no movement when the container is held horizontally for five seconds. |

| | |
|---|---|
| **Pour Point Depressant** | An additive which retards the adverse effects of wax crystallization, and lowers the pour point. |
| **Power Converter** | Devices that are used for converting power from DC to AC, or vice versa. They are used in motor controllers, chemical processing, energy storage and large industrial processes. |
| **Power Factor (pf)** | The ratio of the average (or active) power to the apparent power (root-mean-square voltage times the rms current) of an alternating-current circuit. Synonymous with *phase factor*. |
| **Power Spectral Density (PSD)** | Describes the power of random vibration intensity, in mean-square acceleration per frequency unit, as $g^2/Hz$ or $m^2/s^3$. |
| **Precedence Diagramming Method (PDM)** | A schedule network diagramming technique in which schedule activities are represented by boxes (or nodes). Schedule activities are graphically linked by one or more logical relationships to show the sequence in which the activities are to be performed. |
| **Precision** | The aspect of measurement that addresses repeatability, or consistency, when an identical item is measured several times. |
| **Predictive Maintenance (PdM)** | An equipment maintenance strategy based on measuring the condition of equipment in order to assess whether it will fail during some future period, and then taking appropriate action to avoid the consequences of that failure. The condition of equipment could be measured using condition monitoring, statistical process control, equipment performance, or through the use of the human senses. Synonymous with *condition based maintenance* and *on-condition maintenance*. |
| **Pre-Engineering Phase** | Preliminary studies outlining the basic assumptions, general approach, and probable budget and timetable for a proposed facilities construction project. Such studies include process, feasibility, and preliminary engineering studies and comprise the data necessary for preparation of the appropriation request. |

| | |
|---|---|
| **Prequalification** | The determination of a contractor's or vendor's ability to deliver a product or service that meets specifications prior to solicitation. |
| **Present Value** | The sum of a stream of future cash flows discounted at a given interest (discount) rate. Synonymous with *present worth*. |
| **Presentation** | Formal spoken reports. They communicate to others what has been done or what is planned, with the purpose of asking for something (support, approval, resources) or helping the audience do something similar. Usually, visual images or props are used to clarify and enhance the message. |
| **Pressure** | Force per unit area, usually expressed in pounds per square inch or pascals (SI). |
| **Pressure Drop** | Resistance to flow created by the element (media) in a filter. Defined as the difference in pressure upstream (inlet side of the filter) and downstream (outlet side of the filter). |
| **Pressure Gauge** | A device that responds to the difference between atmospheric pressure and the pressure in a closed system, such as a tank or pipeline. |
| **Pressure Line Filter** | A filter located in a line conducting working fluid to a working device or devices. |
| **Pressure Vessel** | An enclosed container in which a pressure greater than atmospheric pressure can be maintained. |
| **Preventative Action** | Action taken to remove or improve a process to prevent potential future occurrences of a nonconformance. |
| **Prevention Cost** | The cost incurred by actions taken to prevent a nonconformance from occurring. |
| **Preventive Maintenance (PM)** | An equipment maintenance strategy based on replacing, or restoring, an asset at a fixed interval regardless of its condition. Scheduled restoration tasks and replacement tasks are examples of preventive maintenance tasks. |

|  |  |
|---|---|
| **Preventive Maintenance Costs** | The labor, material, and services cost by company personnel or contractors for work performed as preventive maintenance (PM). This includes operator costs if operators are performing preventive maintenance tasks and these costs are included in total maintenance costs. |
| **Preventive Maintenance Man Hours** | The man-hours of labor by company personnel or contractors for work performed as preventive maintenance (PM). Includes operator labor if all operator maintenance costs are included in total maintenance costs. |
| **Primary Effect** | The immediate effect of the failure on the item being analyzed. |
| **Primary Function** | The function, or functions, which constitutes the main reason why a physical asset or system is acquired by its owner or user. |
| **Primary Winding** | The winding of a motor, transformer, or other electrical device which is connected to the power source. |
| **Principles** | Logic, common sense, proven procedures, or essential rules on which plant operation must be based. |
| **Printed Circuit Board (PCB)** | A flat board made of non-conducting material, such as plastic or fiberglass, on which chips and other electronic components are mounted, usually in predrilled holes designed to hold them. |
| **Prioritization Matrix** | A tool used to choose among several options that have many useful benefits but where not all of them are of equal value. The choices are prioritized according to known weighted criteria and then narrowed down to the most desirable to accomplish the task effectively. |
| **Priority** | The relative importance of a single job in relation to other jobs, operational needs, safety, equipment condition, and the time within which the job should be done. |
| **Proaction** | Any activity that will improve operations, prevent mechanical, process or human failure, or lessen the consequences of failure. |
| **Proactive** | Actions that are planned, scheduled, and executed before a break-down occurs. Includes maintenance prevention activities. |

**Proactive Maintenance**  Maintenance work that is completed to avoid failures or to identify defects that could lead to failures (failure finding). It includes routine preventive and predictive maintenance activities and work tasks identified from them.

**Proactive Tasks**  Tasks undertaken in order to prevent an item from failing. They embrace what is traditionally known as predictive and preventive maintenance. RCM uses the terms scheduled restoration, scheduled discard, and on-condition maintenance.

**Proactive Work**  The sum of all maintenance work that is completed to avoid failures or to identify defects that could lead to failures (failure finding). It includes routine preventive and predictive maintenance activities and work tasks identified from them.

**Probabilistic Risk Assessment**  A "top-down" approach used to apportion risk to individual areas of plant and equipment, and possibly to individual assets, so as to achieve an overall target level of risk for a plant, site, or organization. These levels of risk are then used in risk-based techniques, such as Reliability Centered Maintenance and HAZOP, to assist in the development of appropriate equipment maintenance strategies and to identify required equipment modifications.

**Probability**  A term referring to the likelihood of occurrence of an event, action, or item within a defined time interval.

**Probability Distribution**  The listing of all possible events or outcomes associated with a course of action, and their probabilities. Widely known probability distributions include the binomial distribution and the normal distribution.

**Probe**  An assembly that secures the sensing elements of an instrument. It is inserted into an environment to take readings, allowing easy replacement or service of the equipment.

| | |
|---|---|
| **Problem Solving** | The process of:<br>• Defining a problem<br>• Determining the cause of the problem<br>• Identifying, prioritizing and selecting alternatives for a solution<br>• Implementing a solution |
| **Problem Statement** | A succinct statement that describes what the problem is, where and when it occurs, potential reasons why it occurs, how the process operates at the point of the problem, and who is involved in the problem or in a potential solution. |
| **Procedure** | A series of steps followed in a regular, definitive order to accomplish something. |
| **Procedure Manual** | A formal organization and indexing of a company's procedures. |
| **Process** | A series of interrelated steps, consisting of resources and work activities, which transform inputs into outputs and work together to a common end. A process may or may not add value. |
| **Process Average Quality** | Expected or average value of process quality. |
| **Process Capability** | A statistical measure of the inherent process variability for a given characteristic. A process is said to be "capable" when the output of the process always conforms to the process specifications. |
| **Process Capability Index (Cpk)** | An index used to measure the performance of a process with a non-centered distribution. Synonymous with *capability performance index*. |
| **Process Control** | The methodology for keeping a process within boundaries. Minimizing the variation of a process. |
| **Process Decision Program Charts (PDPC)** | A variant of tree diagrams, a PDPC can be used as a simple alternative to a failure modes and effects analysis (FMEA). |
| **Process Flow Diagram (PFD)** | A depiction of the flow of materials through a process, including rework or repair operations. Synonymous with *process flow chart*. |

| | |
|---|---|
| **Process FMEA** | A failure modes and effects analysis (FMEA) that studies how failure in the manufacturing or service process affects the system operation. |
| **Process Hazard Analysis (PHA)** | A set of organized and systematic assessments of the potential hazards associated with an industrial process and ways to manage them. Synonymous with *process hazard evaluation*. |
| **Process Improvement** | The application of the plan-do-study-act (PDSA) philosophy to processes to produce positive improvement and meet the needs and expectations of customers. |
| **Process Improvement Team (PIT)** | A natural work group comprised of cross-functional team members whose responsibility it is to achieve needed improvements in existing processes. |
| **Process Kaizen** | Improvements made at an individual process or in a specific area. Synonymous with *point kaizen*. |
| **Process Management** | The pertinent techniques and tools applied to a process to implement and improve process effectiveness, hold the gains, and ensure process integrity in fulfilling customer requirements. |
| **Process Manufacturing** | The manufacture of products such as chemicals, gasoline, beverages, and food products that typically are produced in "batch" quantities rather than discrete units. Many process operations require inputs such as heat, pressure, and time (for thermal or chemical conversion). |
| **Process Map** | A type of flowchart depicting the steps in a process, with identification of responsibility for each step and the key measures. |
| **Process Mapping** | A technique for indicating flows or steps in a process using standard symbols. Used to facilitate process improvements. |
| **Process Owner** | The person who coordinates the various functions and work activities at all levels of a process, has the authority or ability to make changes in the process as required, and manages the entire process cycle to ensure performance effectiveness. |
| **Process Performance Index ($C_{pk}$)** | A capability index that takes into account the mean. |

| | |
|---|---|
| **Process Quality** | The value of percentage defective, or of defects per hundred units, in product from a given process. Note: The symbols "p" and "c" are commonly used to represent the true process average in fraction defective or defects per unit, and "l00p" and "100c" the true process average in percentage defective or in defects per hundred units. |
| **Process Reengineering** | A strategy directed toward major rethinking and restructuring of a process. Often referred to as the "clean sheet of paper" approach. |
| **Process Safety Management (PSM)** | A comprehensive management program that integrates technologies, procedures, and management practices to manage hazards associated with highly hazardous chemicals. PSM is prescribed by OSHA in 29 CFR 1910.119. |
| **Process Simulation** | The use of a mathematical model by a computer program to envision process design scenarios with real-time visual and numerical feedback. Process optimization and the ability to forecast potential problems are the results. |
| **Procurement Cycle Time** | The total elapsed time from the initiation of a parts requisition (manually or computer generated) until receipt of the part on-site. |
| **Procurement Reliability Specifications** | The specifications used in the procurement of equipment to define the minimum design and performance standards to ensure it meets expected reliability needs. Common examples are pre-defined alignment and balance requirements for rotating equipment, minimum MTBF, MTTR, requirements, etc. |
| **Procurement Cycle Time** | The total elapsed time from the initiation of a parts requisition (manually or computer generated) until receipt of the part on-site. |
| **Product Development Cycle** | The period of time from the start of design/development work to commercial product availability. Synonymous with *time to market*. |
| **Product Life Cycle** | The total period of time that a product exists in the marketplace, from concept to termination. |
| **Product Quality** | The conformance of a product to requirements, specifications, and standards of quality. |

| | |
|---|---|
| **Product Warranty** | An organization's stated policy that it will replace, repair, or reimburse a buyer for a product in the event a product defect occurs under certain conditions and within a stated period of time. |
| **Production Planner** | A person responsible for determining production details and timelines. |
| **Production Strategy** | A plan to produce output with minimum cost. |
| **Production Support Center** | A controlled area authorized for maintaining and stocking specialized tooling/equipment, locally manufactured, modified, or specialized end item, unique tools and equipment. |
| **Production Workers** | Employees directly involved in manufacturing or operational processes, as distinguished from supervisory, sales, executive, and office employees. |
| **Productive Work Time** | The percentage of time an employee spends applying physical effort or attention to a tool, equipment, or materials in the accomplishment of assigned work. It is used to determine how efficient the plant is at planning, scheduling and executing work. |
| **Productivity** | The relationship between the amount or volume of output or service provided and the resources contributing to the actual production of the output or services. Typically, it is the ratio of actual production to standard production, applicable to either an individual worker or group of workers, expressed as units per man-hour or man-hours per unit. |
| **Profit Center** | A unit within a large company that is held responsible for minimizing costs and maximizing revenue. |
| **Prognosis** | The ability to predict or forecast the future condition of a component, or system of components, in terms of either failure or degraded condition, so that it can satisfactorily conform to operational requirements. |
| **Prognostics** | The ability to predict or forecast the future condition of an asset or component. |

**Program – Computer**  A sequence of instructions that can be executed by a computer. The term can refer to the original source code or to the executable (machine language) version. Synonymous with *software*.

**Program – Project**  A group of related projects managed in a coordinated way to obtain benefits and control not available from managing them individually. Programs may include elements of related work outside of the scope of the discrete projects in the program.

**Program Evaluation and Review Technique (PERT)**  An event-oriented, probability-based network analysis technique used to estimate project duration when there is a high degree of uncertainty with the individual activity duration estimates. PERT applies the critical path method to a weighted average duration estimate.

**Program Management**  The centralized, coordinated management of a program to achieve the program's strategic objectives and benefits.

**Programmable Logic Controller (PLC)**  A digital micro computer used for automation of electromechanical processes.

**Programming Language**  Any artificial language that can be used to define a sequence of instructions that can ultimately be processed and executed by the computer.

**Project**  A temporary undertaking to create a unique product or service with a defined start and end point and specific objectives that, when attained, signifies completion.

**Project Charter**  A document issued by the project initiator or sponsor that formally authorizes the existence of a project, and provides the project manager with the authority to apply organizational resources to project activities.

**Project Closeout**  The process to provide for project acceptance by the pro sponsor, completion of various project records, final revision and issue of documentation to reflect the "as-built" condition, and retention of essential project documentation.

**Project Management**  The application of knowledge, skills, tools, and techniques to project activities to meet project requirements.

| | |
|---|---|
| **Project Management Professional (PMP)** | A certification awarded by the Project Management Institute (PMI) to a person, who demonstrates knowledge and skills in leading and directing project teams and delivering project results within the constraint of the schedule, after successfully passing a written test. |
| **Project Plan** | A formal, approved document, in summarized or detailed form, used to guide both project execution and control. It documents planning assumptions and decisions, facilitates communication among stakeholders, and documents approved scope, cost, and schedule baselines. Synonymous with *project manual*. |
| **Project Scope** | The work that must be performed to deliver a product, service, or result with the specified features and functions. |
| **Project Team** | A team that manages the work of a project. The work typically involves balancing competing demands for project scope, time, cost, risk and quality, satisfying stakeholders with differing needs and expectations, and meeting identified requirements. |
| **Project Work** | Actions such as construction, equipment modification, installation, or relocation to gain economic advantage, replace worn, damaged or obsolete equipment, satisfy a safety requirement, attain additional operating capacity, or meet a basic need. |
| **Propagation** | The advancement of energy through a medium. |
| **Proportional Band** | The range of measured values needed to cause maximum possible change in the final control element setting, i.e., the amount of pen movement necessary to give full value movement. Usually expressed as a percent of full scale range. |
| **Proportional plus Integral plus Derivative (PID) Control** | A control that is used in processes where the controlled variable is affected by long lag times. |
| **Pros and Cons** | Arguments for or against a particular issue. Pros are arguments which aim to promote the issue, while cons suggest points against it. Used to evaluate a set of options. |

| | |
|---|---|
| **Protective Device** | Devices and assets intended to eliminate or reduce the consequences of equipment failure. Some examples include standby plant and equipment, emergency systems, safety valves, alarms, trip devices, and guards. |
| **Protocol – Computer** | A standard set of procedures to allow data to be transferred among systems. |
| **Proximity Sensor** | Usually a displacement sensor for measuring the varying distance between a housing and a rotating shaft. |
| **Proximity Switch** | A device that senses the presence or absence of an object without physical contact and that, in response, closes or opens circuit contacts. |
| **Psychometric Chart** | A graph showing the relationships among dew point, relative humidity, and temperature of air. |
| **Pugh Matrix** | A tool that helps determines which potential solutions are more desirable than others. All solutions are evaluated in terms of their strengths and weaknesses and then assigned scores. This tool is usually associated with the quality functional deployment (QFD) method. Synonymous with *decision matrix*. |
| **Pull Production** | The opposite of push production. It means products are made only when the customer has requested or "pulled" it, and not before. Prevents building products that are not needed. |
| **Pull System** | The opposite of "push" system. In "pull" system, the customer process withdraws the item needed from inventory, and the supplying process produces to replenish what was withdrawn. |
| **Pump** | A device for raising, compressing or transferring fluids. |
| **Pumpability** | The low temperature, low shear stress-shear rate viscosity characteristics of an oil that permit satisfactory flow to and from the engine oil pump and subsequent lubrication of moving components. |
| **Purchase Order (PO)** | The prime document created by an organization, and issued to an external supplier, ordering specific materials, parts, supplies, equipment or services. |

**Purchase Requisition**   The prime document raised by user departments authorizing the purchase of specific materials, parts, supplies, equipment or services from external suppliers.

**Pyroelectric Detector**   A type of thermal infrared detector that acts as a current source with its output proportional to the rate of change of its temperature.

**Pyrometry**   The measurement of fire or hot objects, such as the monitoring of furnace or foundry conditions. The measuring instrument used is called a pyrometer.

# Q

**Q9000 series** — Refers to ANSI/ISO/ASQ Q9000 series of standards, which is the verbatim American adoption of the 2000 edition of the ISO 9000 series standards.

**QEDS Standards Group** — The U.S. Standards Group on Quality, Environment, Dependability and Statistics (QEDS), which consists of the members and leadership of organizations who are concerned with the development and effective use of generic and sector specific standards on quality control, assurance and management, environmental management systems and auditing, dependability and the application of statistical methods.

**QS-9000** — A quality management standard developed by the Big Three Automakers for the automotive sector. Currently largely replaced by Technical Specification 16949 (ISO/TS 16949). Quality standards for the United States automotive industry (similar to the ISO-9000 standards).

**Quadrature Motion** — Any motion perpendicular to the reference axis. Shakers are supposed to have zero quadrature motion. Synonymous with *side motion, lateral motion,* and *cross-talk.*

**Quadrature Sensitivity** — A vibration sensor's sensitivity to motion perpendicular to the sensor's principal axis. Commonly expressed in percent of principal axis sensitivity. Synonymous with *cross-talk sensitivity.*

**Qualifications-Based Selection (QBS)** — A negotiated, competitive procurement process for selection based on qualifications and competence in relation to the work to be performed.

**Qualified** — The state of having met conditions and requirements set. Could apply to a process or a person.

**Qualified Person** — A person who, because of their knowledge, training, qualifications, certification, or experience, is competent to perform the duties of their job.

**Quality** — The degree to which a set of inherent characteristics fulfills requirements.

**Quality Assessment**  The operational techniques and activities used to evaluate the quality of processes, practices, programs, and services.

**Quality Assurance (QA)**  All the planned and systematic activities implemented within a quality system that can be demonstrated to provide confidence a product or service will fulfill requirements for quality.

**Quality Audit**  A systematic, independent examination and review to determine whether quality activities and related results comply with planned arrangements, and whether these arrangements are implemented effectively and are suitable to achieve the objectives.

**Quality Circles**  Quality improvement or self-improvement study groups composed of a small number of employees (10 or fewer) and their supervisor. Quality circles originated in Japan, where they are called quality control circles.

**Quality Control (QC)**  The operational techniques and activities that sustain a quality of product or service that will satisfy given needs.

**Quality Culture**  Employee beliefs, habits, practices, and traditions concerning quality.

**Quality Engineering**  The analysis of a manufacturing system at all stages to maximize the quality of the process itself, and the products it produces.

**Quality Function Deployment (QFD)**  A structured method in which customer requirements are translated into appropriate technical requirements for each stage of product development and production. The QFD process is often referred to as listening to the voice of the customer.

**Quality Loss Function**  A parabolic approximation of the quality loss that occurs when a quality characteristic deviates from its target value. The quality loss function is expressed in monetary units. The cost of deviating from the target increases quadratically the further the quality characteristic moves from the target. The formula used to compute the quality loss function depends on the type of quality characteristic being used.

**Quality Management (QM)** — The application of a quality management system in managing a process to achieve maximum customer satisfaction, at the lowest overall cost to the organization, while continuing to improve the process.

**Quality Management System (QMS)** — A formalized system that documents the structure, responsibilities, and procedures required to achieve effective quality management.

**Quality of Conformance** — The level of effectiveness of the design and production functions in effecting the product manufacturing requirements and process specifications, while meeting process control limits, product tolerances, and production targets.

**Quality of Design** — The level of effectiveness of the design function in determining a product's operational requirements (and their incorporation into design requirements) that can be converted into a finished product in a production process.

**Quality Plan** — A document, or set of documents, that describe the standards, quality practices, resources and processes pertinent to a specific product, service or project.

**Quality Policy** — An organization's general statement of its beliefs about quality, how quality will come about, and the expected results.

**Quality Rate** — The degree to which product characteristics agree with the requirements specified on the product or output.

**Quality Score Chart** — A control chart for evaluating the stability of a process. The quality score is the weighted sum of the count of events of various classifications in which each classification is assigned a weight.

**Quality Standard** — A framework for achieving a recognized level of quality within an organization. Achievement of a quality standard demonstrates that an organization has met the requirements laid out by a certifying body. Quality standards recognized on an international basis include ISO 9000 and ISO 14000.

**Quality Strategy**  An organizational "game plan" for quality that typically includes a vision statement, a mission statement, goals, and objectives. It should be integrated with an organization's overall business strategy.

**Quality System**  The organizational structure, procedures, processes, and resources needed to implement quality management.

**Quality Tool**  An instrument or technique to support and improve the activities of process quality management and improvement.

**Quality Trilogy**  A three-pronged approach to managing for quality. The three legs are:

1. Quality Planning – developing the products and processes required to meet customer needs.
2. Quality Control – meeting product and process goals.
3. Quality Improvement – achieving unprecedented levels of performance.

**Quality Values**  The principles and beliefs that guide an organization and its people toward the accomplishment of its vision, mission, and quality goals.

**Quantitative Data**  A term usually used to describe data in which the variables concerned are quantities, as distinct from data derived from qualitative attributes.

**Quasi-Periodic**  A quasi-periodic signal is a deterministic signal whose spectrum is not a harmonic series but nevertheless exists at discrete frequencies. The vibration signal of a machine that has non-synchronous components resembles a quasi-periodic signal. In most cases, a quasi-periodic signal actually is a signal containing two or more different periodic components.

**Query – Computer**  A specific set of instructions for extracting particular data.

**Queuing Theory**  Mathematical modeling of waiting lines, whether of people, signals, or things. It aims to estimate if the available resources will suffice in meeting the anticipated demand over a given period.

**Quick-Changeover Methods** A variety of techniques, such as single-minute exchange of dies (SMED), which reduce equipment setup time and permit more frequent setups, thus improving flexibility and reducing lot sizes and lead times.

# R

| | |
|---|---|
| **R-Value** | The measure of a materials thermal resistance. It is defined as the inverse of thermal conductivity. |
| **Races** | The surfaces on the cup and cone of a bearing where the rolling elements make contact. |
| **Raceway – Bearing** | The functional surfaces in an anti-friction bearing that contact the rolling elements. |
| **Raceway – Electrical** | A channel for loosely holding wires or cables in interior work that is designed expressly and used solely for this purpose. |
| **Radar Chart** | A graphical method of displaying multivariate data in the form of a two-dimensional chart of three or more quantitative variables represented on axes starting from the same point. |
| **Radial** | A direction perpendicular to a shaft's centerline. |
| **Radial Load** | A load applied perpendicular to the axis of the shaft. |
| **Radiant Energy** | Energy transmitted through a medium by electromagnetic waves. |
| **Radiant Flux** | Radiant energy's rate of flow, measured in watts. |
| **Radiation** | Particles or waves emitted from a material. In thermography, radiation relates to heat emitted from a surface. |
| **Radio Frequency (RF) Identification** | An electronic method of assigning a piece of information to a product, process, or person with a transponder (tag) and reading the information with a reader. |
| **Radiometric** | A non-contact temperature measurement based on the thermal radiation emitted by a surface. |

| | |
|---|---|
| **Radiosity** | The total infrared energy (radiant flux) leaving a target surface. It is composed of radiated, reflected, and transmitted components. |
| **Random Access Memory (RAM)** | Semiconductor-based memory that can be read and written by the central processing unit (CPU) or other hardware devices. |
| **Random Cause** | A cause of variation due to chance and not assignable to any factor. |
| **Random Experiment** | A type of experiment where the outcomes or observations in any one trial of such an experiment cannot be predicted with certainty. |
| **Random Failure** | A failure whose occurrence is predictable only in a probabilistic or statistical sense. |
| **Random Measurement Error** | The result of a measurement minus the mean that would result from an infinite number of measurements of the same measurand carried out under repeatability conditions. Synonymous with *precision*. |
| **Random Number** | A number which is non-repetitive and satisfies no algorithm. Random numbers provide a way of selecting a sample without human bias. |
| **Random Number Generator** | A program that generates a sequence of numbers that seem to be completely random. |
| **Random Sampling** | A commonly used sampling technique in which sample units are selected so that all combinations of n units under consideration have an equal chance of being selected as the sample. |
| **Random Vibration** | A vibration whose instantaneous magnitudes cannot be predicted. It may be broad-band, covering a wide, continuous frequency range, or narrow band, covering a relatively narrow frequency range, and has no periodic or deterministic components. |
| **Range** | The measure of dispersion in a data set (the difference between the highest and lowest values). Also the set of values within which measurements can be made without changing the sensitivity of the measuring instrument. |

**Range Chart** — A control in which the subgroup range is used to evaluate the stability of the variability within a process. A plot of the range for each sample with calculated control limits. Synonymous with *R chart*.

**Rankin Scale** — An absolute temperature scale with degrees equivalent to the Fahrenheit scale.

**Ranking Matrix** — A tool used to evaluate and prioritize a list of options by means of ranking. Synonymous with *decision matrix*.

**Rapid Prototyping** — A variety of techniques for quick conversion of CAD-generated product designs into useful, accurate, physical models, typically using computer-controlled systems.

**Raster** — In the display on a video display or raster graphics screen, the grid pattern of vertical and horizontal divisions outlining all the small elements of which the picture is composed.

**Raster Graphics** — A method of generating graphics that treats an image as a collection of small, independently controlled dots (pixels) arranged in rows and columns.

**Rate Gyro** — A kind of gyroscope that measures rotational velocity (degrees or radians per second) around a fixed axis.

**Rate of Return** — The ratio expressing the relationship between the amount of an investment and the income it will produce.

**Rate of Shear** — The difference between the velocities along the parallel faces of a fluid element divided by the distance between the faces.

**Rated Capacity** — The expected output capability of a resource or system. Capacity is traditionally calculated from such data as planned hours, efficiency, and utilization.

*Rated Capacity = hours available x efficiency x utilization*

**Rated Pressure** — The pressure which is recommended for a component or system by the manufacturer as a safe pressure for operation.

**Ratio of Replacement Asset Value to Craft/Wage Head Count** — The replacement asset value (RAV) of the assets being maintained at the plant, divided by the craft/wage employee head count. The result is expressed as a ratio in dollars per person.

**Rattle**  A sound exemplified by shaking a steel can full of steel nuts and bolts.

**Rayleigh Curve**  A roughly bell-shaped curve that represents the buildup and decline of staff power, effort, or cost, followed by a long trail representing staff power, effort, or cost devoted to enhancement or maintenance.

**Rayleigh Distribution**  A special case of the Weibull failure distribution where the Beta (slope) value is known to equal 2.

**Reactance**  The characteristic of a coil in an induction motor when connected to alternating current, which causes the current to lag the voltage in time phase. The current wave reaches its peak later than the voltage wave reaches its peak.

**Reaction Time**  The time to react to correct or solve a problem.

**Reactive Maintenance**  Maintenance repair work done as an immediate response to failure events normally without planning, always unscheduled. Synonymous with *breakdown* and *emergency maintenance.*

**Reactive Work**  Maintenance work that breaks into the weekly schedule, including emergency work (to correct safety concerns, immediate hazards, etc.).

**Read-Only Memory (ROM)**  Any semiconductor circuit serving as a memory that contains instructions or data that can be read but not modified.

**Ready Backlog**  The quantity of work that has been fully prepared for execution but which has not yet been executed. It is work for which all planning has been done and materials procured but is awaiting assigned labor for execution.

**Ready Line**  Equipment, usually mobile, which is available but not being utilized. It is typically parked on the ready line.

**Real Property**  Land, buildings, structures, and utility distribution systems.

**Realbody**  An object that radiates less energy than a blackbody, at the same temperature, however, emitted energy varies with wave length.

**Real-Time Closed Loop Control**  Resembles iterative closed loop control but continuously modifies drive signals throughout the test.

**Real-Time Feedback**  Instantaneous (or nearly instantaneous) communication of electronically captured data (typically quality data), to process operators or equipment, to enable rapid or automated adjustments that keep production processes operating within quality parameters.

**Real-Time System**  Computers designed to receive, process, and respond to data within a time frame set by outside events, e.g., for air traffic control. A system consists of a controlling system and a controlled system. A controlling system interacts with its environment based on information from various sensors and inputs. In many real-time systems, severe consequences result if the timing and logical correctness of the system are not satisfied.

**Rebaselining**  Establishing a new project baseline because of sweeping or significant changes in the project scope.

**Rebuild**  To restore an item to an acceptable condition in accordance with the original design specifications.

**Record**  A document of an event or activity (e.g. equipment history, work order).

**Record – Computer**  A data structure that is a collection of fields (elements), each with its own name and type.

**Record Retention**  The period of time that records are kept for reference purposes. Usually set by company policy or legal requirement.

**Recordable Occupational Injury or Illness**  Every occupational death, illness, or injury which involves days away from work, restricted work, or work transfer. A log of recordable injuries and illnesses must be available for OSHA inspection. Synonymous with *OSHA recordable incident*.

**Records Management**  Procedures established by an organization to identify, index, archive and control distribution of project documents.

**Red Herring** — An argument, given in reply, that does not address the original issue. It is a deliberate attempt to change the subject or divert the argument.

**Redesign** — Any one-off intervention to enhance the capability of a piece of equipment, a job procedure, a management system or people's skill.

**Reducer** — A connector having a smaller line size at one end than the other.

**Redundancy** — The spare capacity which exists in a given system which enables it to tolerate failure of individual equipment items without total loss of function over an extended period of time.

**Reengineering** — A fundamental rethinking and redesign to achieve dramatic improvements in business process performance.

**Reference Mark** — A diagnostic point or mark which can be used to relate a position during rotation of a part to its location when stopped. Synonymous with *hash mark*.

**Reference Plane** — A plane perpendicular to a shaft axis to which a degree of unbalancing is referred.

**Referent Authority** — Influence based on an individual's referring to a higher power as supporting his or her position or recommendation. ("The boss and I think this is a good idea.")

**Reflective Listening** — Listening behaviors that provide feedback that the message was communicated (e.g., paraphrasing, clarifying, summarizing).

**Reflectivity** — The ratio of the intensity of the total energy reflected from a surface to total radiation on the surface.

**Refraction** — The change of direction or speed of light as it passes from one medium to another.

**Refurbishment** — Extensive work intended to restore a piece of equipment to acceptable operating condition.

**Registrar Accreditation Board (RAB)**    A board that evaluates the competency and reliability of registrars (organizations that assess and register companies to the appropriate ISO 9000 series standards and to the ISO 14000 environmental management standard). RAB provides ISO course provider accreditation. Formed in 1989, RAB is governed by a board of directors from industry, academia, and quality management consulting firms and by a joint oversight board for those programs operated with the American National Standards Institute.

**Registration**    The act of including an organization, product, service, or process in a compilation of those having the same or similar attributes.

**Registration to Standards**    A process in which an accredited, independent, third-party organization conducts an on-site audit of a company's operations against the requirements of the standard to which the company wants to be registered. Upon successful completion of the audit, the company receives a certificate indicating that it has met the standard requirements.

**Regression**    The relationship between the mean value of a random variable and the corresponding values of one or more independent variables.

**Regression Analysis**    A statistical technique for determining the best mathematical expression describing the functional relationship between one response and one or more independent variables.

**Regulation**    A rule, ordinance, or law by which conduct is regulated.

**Rejection Number**    The smallest number of defectives (or defects) in the sample or samples under consideration that will require the rejection of the lot.

**Relationship Map**    A graphical technique that provides an organized visual display of concepts, related vocabulary, and background knowledge of a subject.

**Relative Humidity**    The ratio (in percent) of the water vapor content in the air to the maximum content possible at that temperature and pressure.

| | |
|---|---|
| **Relay** | An electromechanical device for remote or automatic control that is actuated by variation in conditions of an electric current. The relay operates other devices in the same, or a different circuit. |
| **Relevant Failure** | A failure which can occur or recur during the operational life of an item. Inventory failures which occur during development and testing may be categorized as relevant or non-relevant to the end use of the system. |
| **Reliability** | The probability that equipment, machinery, or systems will perform their required functions satisfactorily under specific conditions within a certain time period. |
| **Reliability Allocation** | A step in the design process to translate the overall system reliability requirements into reliability requirements for each of the subsystem or components. Synonymous with *reliability apportionment* and *reliability appraisal*. |
| **Reliability Analysis** | The process of identifying maintenance of significant items and classifying them with respect to malfunction on safety, environmental, operational, and economic consequences. The possible failure mode of an item is identified and an appropriate maintenance policy is assigned to counter it. Support methods include failure mode, effect, and criticality analysis (FMECA), fault tree analysis (FTA), risk analysis, and hazardous operations (HAZOP) analysis. |
| **Reliability and Maintainability Engineering** | A function intended to ensure that maintenance techniques are effective, equipment is designed for optimum maintainability, persistent and chronic problems are analyzed, and corrective actions or modifications are implemented. |
| **Reliability and Maintenance Engineer** | The person specifying, analyzing, and directing operation and maintenance actions during the life cycle of an asset. |
| **Reliability Assurance** | Actions necessary to provide adequate confidence that the product or equipment conforms to established reliability requirements. |

| | |
|---|---|
| **Reliability Block Diagram (RBD)** | A visual representation of a system that contains redundancy in its elements. Redundancy is used for mission critical functions where a single-point failure is not acceptable and reliability needs to be improved. Generally represented in a graphical manner, a simple RBD model may be calculated using analytical solutions, while a more complex RBD may be calculated using a Monte Carlo simulation. |
| **Reliability Centered Maintenance (RCM)** | A structured process, originally developed in the airline industry, but now commonly used in all industries to determine the equipment maintenance strategies required for any physical asset to ensure that it continues to fulfill its intended functions in its present operating context. |
| **Reliability Data** | The data necessary to calculate reliability numbers. For example, calculating MTBF requires the number of failures and asset operating time. |
| **Reliability Deficiency Analysis** | An evaluation of asset performance to determine the optimum improvement course for reliability improvement. The evaluation is normally based on a review of documented discrepancies and data resulting from corrective and preventive actions, and in-service inspections. |
| **Reliability Engineering** | A function whose prime responsibility is to ensure that maintenance techniques are effective, that equipment is designed and modified to improve maintainability, that ongoing maintenance technical problems are investigated, and appropriate corrective and improvement actions are taken. |
| **Reliability Goal** | A goal generally associated with a reliability growth program or reliability improvement program. A program may have more than one reliability goal. For example, there may be a reliability goal associated with failures resulting in failures—unscheduled maintenance actions, and a separate goal associated with those failures causing a mission abort or catastrophic failure. Other reliability goals may be associated with failure modes that are safety related. Reliability growth management addresses the attainment of the reliability objectives through planning and controlling of the reliability improvement process. |

**Reliability Growth** — Machine reliability improvement as a result of identifying and eliminating machinery or equipment failure causes during machine-testing and operation.

**Reliability Indicator** — A measure of equipment, process, or facility ability to meet operating availability and reliability goals. Examples include mean time between failure (MTBF), and overall equipment effectiveness (OEE).

**Reliability Information Systems** — Systems that utilize data collected by CMMS/EAM systems and apply reliability algorithms for the purpose of identifying opportunities to improve asset reliability.

**Reliability Objective** — Sometimes used interchangeably with reliability goal, but usually applied on a broader scale with less finality, as when coordinating sub-goals with higher level organizational goals.

**Reliability Parameter** — A measure of reliability such as mean time between failures, failure rate, probability of survival, or probability of success. The reliability-creating factor of the product development process used as an index of design reliability to facilitate trade-offs in the pursuit of reliability achievement or growth.

**Reliability Prediction** — A primary component of reliability analysis, it is often referred to as the failure rate, or the number of failures expected during a certain period of time. Calculation of equipment failure rate, and the related MTBF (Mean Time Between Failures), is the basis of performing a reliability prediction analysis.

**Reliability Qualification Testing (RQT)** — Testing used to verify that a product, which has been built according to particular design specifications, conforms to the intended design reliability standards. Synonymous with *reliability demonstration* and *designed approval testing*.

**Reliability Software** — Software which helps perform reliability analysis such as FMEA, RBD, and reliability predictions.

**Reliability Target** — A general level of reliability typically proposed in the planning or formative stages of a process or product development program to indicate a desired result when the degree of attainment is still in question.

| | |
|---|---|
| **Relief Valve** | A specialized valve that is held in the closed position by a spring. It opens automatically when pressure in the system exceeds a preset value and closes again automatically when the over-pressure condition has been corrected. |
| **Relocation** | Repositioning major equipment to perform the same function in a new location. |
| **Reluctance** | The resistance offered to magnetic flux by a magnetic circuit, equal to the magnetomotive force divided by the magnetic flux. Similar to the resistance in an electric circuit. |
| **Remote Devices** | Devices that are located a considerable distance from the computer or processing instrument. |
| **Reorder Point** | The inventory level that signals the need to place a new order. |
| **Reorder Quantity** | The quantity which should be ordered each time the available stock (on hand plus on order) falls below the order point in a fixed reorder system. However, in a variable reorder system the amount ordered from time period to time period varies. Synonymous with *replenishment order quantity*. |
| **Repair** | An activity which returns the capability of an asset that has failed to a level of performance equal to, or greater than, that specified by its functions. |
| **Repair Event** | The act of restoring the function of an asset after failure or imminent failure. Synonymous with *corrective work*. |
| **Repair History** | A record of significant repairs made on key equipment used to spot chronic, repetitive problems, failure patterns, and component lifespan which, in turn, identifies corrective actions and helps forecast component replacements. |
| **Repair Parts** | Individual parts or assemblies required for the maintenance or repair of equipment, system, or spares. Such repair parts may be repairable or non-repairable assemblies or one-piece items. Consumable supplies used in maintenance, such as wipe rags, solvent, and lubricants, are not considered repair parts. |

| | |
|---|---|
| **Repair Time** | The time required to restore the function of an asset after failure or imminent failure. It includes both scheduled and unscheduled repair time. |
| **Repairable** | An item that can be restored to perform all of its required functions by corrective maintenance. |
| **Repeatability** | The ability to reproduce a detectable indication in separate processing and tests from a constant source. |
| **Repetitive Maintenance** | Maintenance jobs which have a known labor and material content and occur at a regular interval. |
| **Repetitive Manufacturing** | Production of discrete units, planned and executed via a schedule, usually at relatively high speeds and volumes. Material tends to move in a sequential flow. |
| **Replace** | Removal of an entire component and installation of a new or rebuilt equivalent component. |
| **Replacement Asset Value (RAV)** | The monetary value that would be required to replace the production capability of the present assets in the plant. Includes production or process equipment, as well as utilities, support, and related assets. It should not be based on the insured value or depreciated value of the assets. It includes the replacement value of buildings and grounds if these assets are maintained by the maintenance expenditures. It does not include value of real estate – only improvements. Synonymous with *estimated replacement value (ERV)*. |
| **Replacement Unit** | Any unit that is designed and packaged to be readily removed and replaced in an equipment system without unnecessary calibration or adjustment. |
| **Replenishment** | Re-stocking primary picking locations from reserve storage locations. |
| **Replenishment Lead Time** | The total period of time that elapses from the moment it is determined that a product is to be reordered until the order is available for use. |
| **Replication** | The process of copying a database (or parts of it) to other parts of a network in a distributed database management system. Replication allows distributed database systems to remain synchronized. |

| | |
|---|---|
| **Report** | A written document containing the results of a study or investigation (e.g., work study report) or which summarizes activities during a time period (e.g., monthly report). Its purpose is to inform. |
| **Reprocessing Losses** | Recycling losses resulting from rejected material or product that must be sent to a prior stage in the process to make it acceptable. |
| **Reproducibility** | One of the main principles of the scientific method. It refers to the ability of a manufacturing system, experiment, or test to be accurately reproduced, or replicated, by someone else working independently. |
| | To produce a part, image, or copy of, consistently, without any issues. |
| **Request for Proposal (RFP)** | Type of bid document used to solicit proposals from prospective contractors for products or services. Used when items or services are of a complex nature and assumes that negotiation will take place between the buyer and the contractor. |
| **Request for Quotation (RFQ)** | A document used to solicit proposals from prospective contractors for standard products or services such that negotiation may not be required. Similar to a request for proposals, but generally with a lower monetary amount involved in the procurement. |
| **Request for Stock Code** | A formal written request to establish a stock item necessary for maintenance of an asset. The item is listed on the bill of materials of the CMMS and linked to the asset record. |
| **Required Completion Date** | The required date of completion assigned to a specific activity or project. |
| **Required Operating Time** | The number of hours the production equipment (item) is required to perform a required function by the user. Required operating time includes transition (tool change, required process cleaning) time. |
| **Requirement** | An established, requisite characteristic of a product, process, or service. |
| **Requisition** | The first step of the purchasing cycle resulting from the request for purchases of goods and services from a vendor. |

**Rerefining** — A process of reclaiming used lubricant oils and restoring them to a condition similar to that of virgin stocks by filtration, clay adsorption, or more elaborate methods.

**Rescheduling** — The process of changing the duration or dates of an existing schedule because of unanticipated problems encountered during the task, or due to externally imposed conditions.

**Reservoir** — A container for storage of liquid in a fluid power system.

**Residual Dirt Capacity** — The dirt capacity remaining in a service loaded filter element after use, but before cleaning, measured under the same conditions as the dirt capacity of a new filter element.

**Residual Life** — The remaining useful life of a piece of equipment from the current operating time.

**Residual Unbalance** — The unbalance that remains after balancing.

**Residual Value** — The value of a fixed asset after depreciation charges have been subtracted from its original cost.

**Resistance – Electrical** — An obstruction to the free flow of electrons in a material. It is measured in ohms. Resistance causes electric current to lose energy in the form of heat.

**Resistance – Thermal** — A measure of a material's resistance to the flow of thermal energy, inversely proportional to its thermal conductivity.

**Resistance To Ground (RTG) Testing** — Measuring the leakage current flowing through an insulation system to ground to determine the integrity of the insulation.

**Resolution** — The smallest change in input that will produce a detectable change in an instrument's output. Differs from precision in that human capabilities are involved.

**Resonance** — The tendency of a system to oscillate at larger amplitude at some frequencies more than at others.

**Resource Breakdown Structure (RBS)** — A variation of the organizational breakdown structure used to show which work elements are assigned to individuals.

**Resource Histogram**    Vertical bar chart used to show resource consumption by time period. Synonymous with *resource loading chart*.

**Resource Leveling**    A process of evening out the peaks and valleys of resource requirements so that a fixed amount of resources can be used over time.

**Resource Loading**    Designating the amount and type of resources to be assigned a specific activity in a certain time period.

**Resource Matrix**    Structure used to allocate types of resources to tasks by listing the tasks in a work breakdown structure (WBS) along the vertical axis and the resources required along the horizontal axis. Differs from a responsibility assignment matrix in that assignments of specific individuals are typically not depicted.

**Resource Plan**    Document used to describe the number of resources needed to accomplish work for a project and the steps necessary to obtain a resource.

**Resource Planning**    Process of determining resources (people, equipment, materials) needed in specific quantities, and during specific time periods, to perform activities for a project.

**Resources**    The inputs necessary to carry out an activity (people, money, tools, materials, equipment).

**Response**    A quantitative expression of the output of an instrument (or system) as a function of an input under explicit conditions.

**Response Signal**    The signal from a response sensor measuring the mechanical response of a mechanical system to an input vibration or shock.

**Responsibility**    The obligation of an individual or group to perform assignments effectively.

**Responsibility Assignment Matrix (RAM)**    A structure used to relate a work breakdown structure (WBS) to ensure that each element of the scope of work is assigned to an individual. A high-level RAM defines which group or unit is responsible for each WBS element. A low-level RAM assigns roles and responsibilities for specific activities to particular people. Synonymous with *accountability matrix*.

| | |
|---|---|
| **Restoration** | Actions taken to restore an asset to its desired functional state. |
| **Restructuring** | A method of increasing organizational efficiency and effectiveness that involves making changes in organization structure. Synonymous with *reorganization*. |
| **Results** | The effects that relate to what is obtained by an organization at the conclusion of a time period. |
| **Return Line** | A location in a line conducting fluid from working device to reservoir. |
| **Return Line Filtration** | Filters located upstream of the reservoir but after fluid has passed through the system's output components (cylinders, motors, etc.). |
| **Return On Assets (ROA)** | A measure of a company's financial performance, equal to net income divided by the value of fixed assets. |
| **Return On Capital Employed (ROCE)** | A measure of the returns that an organization is realizing from its capital. Calculated as profit before interest and tax divided by the difference between total assets and current liabilities. The resulting ratio represents the efficiency with which capital is being utilized to generate revenue. |
| **Return On Equity (ROE)** | A measure of how well an organization uses reinvested earnings to generate additional earnings. Equal to a fiscal year's after-tax income divided by book value, expressed as a percentage. It is used as a general indication of the organization's efficiency. In other words, how much profit it is able to generate given the resources provided. |
| **Return On Investment (ROI)** | A financial measure of the return from an investment, usually expressed as a percentage of earnings produced by an asset to the amount invested in the asset. |
| **Return On Net Assets (RONA)** | A measure of a company's financial performance. Equal to net income divided by the sum of fixed assets and net working capital. The higher the return, the better the performance. |

**Return On Quality (ROQ)**  The ratio of increase in profit to the cost of quality improvement programs. This measure provides a basis to decide whether or not a quality improvement project is acceptable and to select a better alternative among competing programs.

**Reversal Technique**  A tool used for improving a product or a service. The opposite of the question you want to ask is asked, and the results are analyzed. Used in reverse brainstorming.

**Reverse Brainstorming**  A problem-solving method that utilizes a combination of brainstorming and reversal techniques. Usually in brainstorming, the question is, "How do we solve this problem?" In a "reverse brainstorming" session, the question becomes, "How could I have possibly caused this problem?"

**Reverse Engineering**  A method of copying a technology (as opposed to starting from scratch) which begins with an existing product or asset and works backward to figure out how it does what it does. When the product or asset's basic principle or core concept is determined, the next step is to reproduce the same results by employing different mechanisms.

**Reverse Logistics**  The process of planning, implementing, and controlling the flow of finished goods from the point-of-consumption to the point-of-origin for the purpose of recapturing value or proper disposal.

**Reverse Osmosis**  A water purification process where water is forced through a semi-permeable membrane under pressure sufficient to overcome osmotic pressure, leaving behind a percentage of dissolved organic, dissolved ionic, and suspended impurities.

**Revision**  A corrected or new version of a document or software.

**Reward and Recognition System**  Formal management actions to promote or reinforce desired behavior. To be effective, such a system should make the link between performance and reward clear, explicit, and achievable and should take in to account individual differences (what is considered a highly motivating reward by one person may not be by another).

**Rework**            The corrective (repair) work done on previously maintained equipment because of maintenance, operations, or material problems that resulted in a premature functional failure of that equipment. The causes of rework may be maintenance, operational or material quality issues. Synonymous with *repeat maintenance*.

**RF Monitoring**     A technique used to detect arcs caused by broken windings in motors and generators.

**RFI Noise**         Disturbances to electrical signal from radio frequency interference (RFI).

**Rheology**          The study of the deformation and flow of matter in terms of stress, strain, temperature, and time.

**Right Sizing**      A process that challenges the complexity of equipment. It examines how equipment fits into an overall vision for how work will flow through the factory. When possible, right sizing favors smaller, dedicated machines rather than large, multipurpose, batch-processing machines.

**Right the First Time**  A term used to convey the concept that it is beneficial and more cost effective to take the necessary steps up front to ensure a product or service meets its requirements than to provide a product or service that will need rework or not meet customer needs. In other words, an organization should engage in defect prevention rather than defect detection.

**Rigid Rotor**       A rotor which operates substantially below its first bending critical speed.

**Ring Lubrication**  A system of lubrication in which the lubricant is supplied to the bearing by an oil ring.

**Ring Sticking**     The freezing of a piston ring in its groove, in a piston engine or reciprocating compressor, due to heavy deposits in the piston ring zone.

**Ringing**           Continued oscillation after an external force or excitation is removed, as after a guitar string is plucked.

**Rings**             Circular metallic elements that ride in the grooves of a piston and provide compression sealing during cycling. Also used to spread oil for lubrication.

| | |
|---|---|
| **Rise Time** | The time required for the output of a transducer to rise from 10% to 90% of its final value as it responds to a step change in the measurand. |
| **Risk** | The potential for the realization of any unwanted, negative consequences of an event. |
| **Risk Analysis** | The analysis of the probability and impact of undesirable events. |
| **Risk Assessment** | Review, examination, and judgment to see whether the identified risks are acceptable according to proposed actions. |
| **Risk Management** | An organized, analytic process identifying what can go wrong, quantifying and assessing associated risks, and implementing the appropriate approach to prevent or handle each identified risk. |
| **Risk Prioritization** | Filtering, grouping, and ranking risks following assessment. |
| **Risk Priority Code (RPC)** | A number code that expresses the degree of risk in terms of severity and probability. Synonymous with *risk priority number*. |
| **Risk Priority Number (RPN)** | A technique used for analyzing the risk associated with potential problems identified during a Failure Mode and Effects Analysis (FMEA). RPN utilizes three rating scales:<br>1. Severity – rates the severity of the potential effect of the failure<br>2. Occurrence – rates the likelihood that the failure will occur<br>3. Detection – rates the likelihood that the problem will be detected before it reaches the end-user/customer<br>*Risk Priority Number = severity x occurrence x detection* |
| **RMS Amplitude** | The root mean square (RMS) of a signal taken during one full cycle. |
| **RMS Responding** | A measurement equal to the root mean square (RMS) value of the input signal for all waveforms within the specified frequency range and crest factor limit. |

**Road Map** — A time-sequenced program for introducing future product or process developments. It indicates what will be developed and when it is targeted for delivery.

**Roadblock** — An impediment to progress, or obstruction, which prevents people, teams, and organizations from meeting objectives.

**Robotics** — The study of the design and use of robots, particularly for their use in manufacturing and related processes.

**Robustness** — The condition of a product or process design that remains relatively stable, with a minimum of variation, even though factors that influence operations or usage, such as environment and wear, are constantly changing.

**Roll** — Rotation about the axis of linear motion.

**Rolled Throughput Yield (RTY)** — The probability of a unit of product passing through an entire process defect-free. This measure is calculated by multiplying together quality yield values at various points in a production process, not just at the end of the line. The purpose is to make problem areas within a process more visible. Synonymous with *multiple-point yield*.

**Roller Bearing** — An antifriction bearing comprising rolling elements in the form of rollers.

**Rollers** — The rolling elements that are located between the cone and cup of a bearing.

**Roll-Off Cleanliness** — The fluid system contamination level at the time of release from an assembly or overhaul line. Fluid system life can be shortened significantly by full-load operation under a high fluid contamination condition for just a few hours. Contaminant implanted and generated during the break-in period can devastate critical components unless removed under controlled operating and high performance filtering conditions.

**Root Cause** — A factor that causes a nonconformance and which should be eliminated through design or process improvement.

**Root Cause Analysis (RCA)** — The process of identifying sources of variation to identify the key sources causing a problem. Eliminating these root causes will have the biggest impact on solving the problem.

| | |
|---|---|
| **Root Cause Failure Analysis (RCFA)** | Analysis used to determine the underlying cause or causes of a failure so that steps can be taken to manage those causes and avoid future occurrences of the failure. |
| **Root Mean Square (RMS)** | The square root of the sum of the squares of a series of numbers. |
| **Rotable** | A component which when it has failed, or is about to fail, is removed from the asset and a replacement component installed in its place. The component that has been removed is then repaired or restored, and placed back in the maintenance store or warehouse, ready for re-issue. |
| **Rotary Pump** | A positive displacement pump used mainly to pump liquids that are either too viscous or too difficult to pick up with suction from a centrifugal pump. |
| **Rotating Bomb Oxidation Test (RBOT)** | A test (ASTM D 2272) used to estimate oxidation stability and remaining useful oil life. |
| **Rotating Magnetic Field** | The force created by the stator, once power is applied to it, which causes the rotor in an electric motor to turn. |
| **Rotation** | The direction in which a shaft turns, either clockwise or counterclockwise. When specifying rotation, also state if viewed from the shaft end or the opposite shaft end. |
| **Rotor** | A rotating body whose journals are supported by bearings. |
| **Rotor – Electric** | The rotating member of an induction motor or generator. |
| **Rotor Influence Check** | A test performed using inductance readings taken off-line at specific, successive rotor positions using higher than operating frequency, low voltage AC signals on all three phases of AC motor measurements. These readings give indications of rotor defects and rotor-stator misalignment caused by shaft, bearing, or shaft support abnormalities. |
| **Route Maintenance** | A form of maintenance wherein a mechanic has an established route through the facility to fix minor problems. The route mechanic is usually very well equipped so he/she can deal with most small problems. Route maintenance and preventive maintenance activity are sometimes combined. |

| | |
|---|---|
| **Routine Maintenance** | Maintenance work of a repetitive nature which is undertaken on a periodic time (or equivalent) basis. Synonymous with *routine repairs*. |
| **RS-232, 422, 423, 449** | The standard electrical interfaces for connecting peripheral devices to computers. |
| **Rule-Based System** | A functional system in which knowledge is stored in the form of simple if-then or condition-action rules. |
| **Run Chart** | A chart showing a line connecting numerous data points collected from a process running over a period of time. |
| **Runout** | Measurable irregularity across a plane surface, such as a disc brake rotor, hub, or wheel assembly. |
| **Run-to-Failure** | A maintenance strategy selected for assets where the cost and impact of failure is less than the cost of preventive actions. |
| **Rust Prevention Test** | A test for determining the ability of an oil to aid in preventing the rusting of ferrous parts in the presence of water. |

# S

**SAE J1739**
A failure modes and effects analysis (FMEA) standard of SAE. This standard introduces the topic of FMEA and gives general guidance in the application of the technique.

**Safe Workplace**
A workplace in which the likelihood of all identifiable undesired events are maintained at an acceptable level.

**Safety**
Actions or steps taken to prevent bodily injury or death. Local, state, and federal (OSHA) laws outline safety requirements for a broad range of activities or situations. Compliance with these regulations often requires specific measures be taken to eliminate or reduce worker's risk of exposure to injury or death.

**Safety Coupling**
A friction coupling adjusted to slip at a predetermined torque, to protect the rest of the system from overload.

**Safety Engineering**
An engineering discipline concerned with the planning, development, improvement, coordination, and evaluation of the safety component of integrated systems of individuals, materials, equipment, and environments to achieve optimum safety effectiveness in terms of both protection of people and protection of property.

**Safety Factor**
The factor used to provide a design margin over the theoretical design capacity to allow for uncertainty in the design process.

**Safety Helmet**
Rigid headgear of varying materials designed to protect the head, not only from impact, but from flying particles and electric shock, or any combination of the three. Safety helmets should meet the requirements of ANSI Standard Z89.1, Protective Headwear for Industrial Workers.

**Safety Improvement Programs**
Practices intended to constantly improve safety within a plant or across a company, including, but not limited to, safety teams, safety awareness programs and communications, safety days, safety training, and setting of continuous-improvement goals targeting safety metrics, such as OSHA incidents or lost-workday rates.

**Safety Lock**  A device designed to prevent injury or accidents. It is commonly used to lock out the electrical energized sources in equipment or machinery operation prior to a maintenance activity.

**Safety Rule**  A rule prescribing safeguarding requirements, personal protective equipment, or safe behavior on the job.

**Safety Valve**  A valve on a still, vapor line, or other pressure vessel, set so that it will permit the emission of fluid when the maximum safe working pressure is reached.

**Salvage Value**  The cost recovered, or could be recovered, from a used property when removed from service, sold, or scrapped.

**Sample**  In acceptance sampling, one or more units of product (or a quantity of material) drawn from a lot for purposes of inspection to reach a decision regarding acceptance of the lot.

**Sample Plan**  Documentation of the sample size or sizes to be used, the sampling frequency, and the associated acceptance and rejection criteria.

**Sample Size**  The number of units in a sample.

**Sample Space**  A set of all possible outcomes of a random experiment.

**Sample Standard Deviation Chart**  A control chart in which the subgroup standard deviation, s, is used to evaluate the stability of the variability within a process. Synonymous with *S chart*.

**Sampling**  The act, process, or technique of selecting a suitable sample for testing. Also, measuring at regular intervals the output or variable of a process to estimate characteristics of the process.

**Sampling at Random**  The process of selecting sample units, as commonly used in acceptance sampling theory, so all units under consideration have the same probability of being selected. Synonymous with *random sampling*.

**Sampling Period**  The time between successive instantaneous samples of the process variable in a discrete time control system.

| | |
|---|---|
| **Sampling Rate** | The frequency with which samples are taken. Sampling can be per unit time or per lot. Synonymous with *sampling frequency*. |
| **Sanitizing** | The act of cleaning the work area. Dirt is often the root cause of premature equipment wear, safety problems, and defects. Synonymous with *shine* (one of the 5's). |
| **Satisfier** | A term used to describe the quality level received by a customer when a product or service meets expectations. |
| **Saturation Temperature** | The temperature at which a material changes state from a liquid to a vapor or vice versa. Synonymous with *boiling point* and *condensing temperature*. |
| **Saybolt Universal Viscosity (SUV)** | The time in seconds required for 60 cubic centimeters of a fluid to flow through the orifice of the standard Saybolt universal viscometer at a given temperature under specified conditions (ASTM Designation D 88). |
| **SCADA** | An acronym for Supervisory Control and Data Acquisition, a system that collects data from various sensors at a factory, plant, or other remote locations and then sends this data to a central computer which manages and controls the data. |
| **Scan** | The examination of data from process sensors. |
| **Scan Time** | The time to completely execute a programmable logic controller (PLC) program once, including an input/output update. |
| **Scatter Diagram** | A graphical technique to analyze the relationship between two variables. Two sets of data are plotted on a graph. The y-axis is used for the variable to be predicted and the x-axis is used for the variable to make the prediction. Synonymous with *scatter plot*. |
| **Scenario** | An account or synopsis of a projected course of action, events or situations. |
| **Schedule** | A time-phased work plan from which job tasks can be assigned to crews and/or individuals. |
| **Schedule Baseline** | An approved project schedule that serves as the basis for measuring and reporting schedule performance. |

| | |
|---|---|
| **Schedule Compliance** | A measure of adherence to a job, work, or project schedule. |
| **Schedule Revision** | Changes to the scheduled start and finish dates in a project schedule. |
| **Schedule Variance** | The difference between the scheduled work or activity and its actual completion. |
| **Scheduled** | The work, task or activity that has been placed on a schedule. |
| **Scheduled Component Replacement** | A maintenance strategy based on predictive maintenance or statistical component failure data that targets component replacement just prior to failure and at an optimum schedule opportunity. |
| **Scheduled Discard** | The replacement of a component with a new component at a specified, predetermined frequency, regardless of the condition of the component at the time of its replacement. |
| **Scheduled Downtime** | The time to do required work on an asset that is on the approved maintenance schedule. |
| **Scheduled Finish Date** | A point in time when work is scheduled to finish on an activity. Synonymous with *planned finish date*. |
| **Scheduled Maintenance** | Maintenance work that has been planned and included on an approved maintenance schedule. |
| **Scheduled Outage** | Downtime that is intended for maintenance, servicing, operational, or other purposes. |
| **Scheduled Overhaul** | A maintenance activity undertaken at a scheduled time interval whose primary purpose is to reduce the number of failures and prevent equipment from reaching the age at which frequent failures will cause substantial loss in performance. Synonymous with *scheduled restoration*. |
| **Scheduled Start Date** | The point in time when work is scheduled to start on an activity. Synonymous with *planned start date*. |
| **Scheduled Work** | Work that has been identified in advance and entered in a schedule so that it may be accomplished in a timely manner based upon its criticality. For example, it may be prioritized and lower priority work may be deferred based upon resource availability and criticality. |

| | |
|---|---|
| **Scheduled Work Order** | A work order that has been planned and included on a maintenance schedule. |
| **Scheduled Work Performed** | The actual number of hours spent on work that has been scheduled. |
| **Scheduled Work Time** | The time in a defined period (1 shift = 8 hours or 480 minutes) during which work is scheduled. |
| **Scheduler** | An individual who schedules work. The role of preparing schedules for work crews by matching available resources to planned and prioritized work orders. |
| **Scheduling** | Fitting tasks into a logical timetable with detailed work planning. |
| **Scheduling Cycle** | The length of time for which scheduling is normally done. |
| **Schematic Diagram** | A graphic illustration showing principles of construction, assembly, or operation without accurate mechanical representations. |
| **Scientific Approach** | A term referring to the intent to find and use the best way to perform tasks to improve quality, productivity, and efficiency. |
| **Scope** | The sum of the products and services to be provided by a project. |
| **Scope Change** | A modification to the agreed-upon project scope as defined by the approved work breakdown structure (WBS). |
| **Scope Creep** | A gradual progressive increase of a project's scope that is not noticed by the project management team or the customer. Occurs when the customer identifies additional, sometimes minor, requirements that may collectively result in a significant scope change and cause cost and schedule overruns. |
| **Scope of Work** | A description of the totality of work to be accomplished, or resources to be supplied, under a contract or agreement. Synonymous with *statement of work*. |
| **Scoping** | Determining the job scope. Studying and defining what work a job requires. Deciding the magnitude of work involved. |

**Scorecard** — An evaluation device, usually in the form of a questionnaire, that specifies the criteria customers will use to rate an organization's performance in satisfying their requirements.

**Scoring** — Distress marks on sliding metallic surfaces in the form of long, distinct scratches in the direction of motion. Scoring is an advanced stage of scuffing.

**Scrap** — Parts or materials wasted in the production process.

**Scrap and Rework Costs** — The cost of parts or materials wasted in the production process, plus the cost of fixing defective products so that they pass final inspection.

**Scrap Rate** — The amount of irreversibly damaged product divided by the amount of total product produced by an asset, usually expressed as a percentage of throughput or output.

**Screening** — A technique used to review, analyze, rank, and select the best alternative for the proposed action.

**Scuffing** — Localized distress marks on sliding metallic surfaces, appearing as a matte-finished area rather than as individual score marks.

**Scuffing Particles** — Large, twisted and discolored metallic particles resulting from scuffing.

**S-Curve** — Graphic display of cumulative costs, labor hours, or other quantities, plotted against time. The curve is flat at the beginning and end and steep in the middle. Generally describes a project that starts slowly, accelerates, and then tapers off.

**Seasonal Maintenance** — Maintenance work carried out at a specific time of year (repair of potholed roads during summer months in northern climates).

**Second Law of Thermodynamics** — Heat cannot flow from a cooler object to a warmer one unless additional work or energy is added.

**Secondary Damage** — Any additional damage to equipment, above and beyond the initial failure mode, that occurs as a direct consequence of the initial failure mode. Synonymous with *secondary effect*.

| | |
|---|---|
| **Secondary Functions** | Those functions which a physical asset or system has to fulfill, apart from its primary function(s), that are needed to fulfill regulatory requirements and those which concern issues such as protection, control, containment, comfort, appearance, energy efficiency and structural integrity. |
| **Secretive Management Style** | A management approach in which the manager is neither open nor outgoing in speech, activity, or purpose, to the detriment of the organization. |
| **Seismic** | A kind of sensor that depends upon the inertia of an internal mass to generate a signal (for example, an accelerometer or velocity pickup). |
| **Self-Contained Breathing Apparatus (SCBA)** | A respiratory protection device that consists of a supply of respirable air, oxygen, or oxygen-generating material worn by the worker. |
| **Self-Directed Work Team (SDWT)** | A type of team structure in which much of the decision making regarding how to handle the team's activities is controlled by the team members themselves. Synonymous with *self-managed* and *natural work teams*. |
| **Self-Induced Vibration** | Vibration that results from conversion of non-oscillatory energy into vibration, as wind exciting telephone wires into mechanical vibration. Synonymous with *self-excited vibration*. |
| **Semantics** | The science dealing with the relationship between signs and symbols (including words), their meaning, and human behavior. |
| **Semiconductor** | Any of a class of solids having higher resistivity than a conductor, but lower resistivity than an insulator. Important semiconductors are:<br>• Silicon<br>• Germanium<br>• Lead Sulphide<br>• Selenium<br>• Silicon Carbide<br>• Gallium Arsenide<br><br>Semiconductor materials are the basis of all integrated circuits. |

**Semi-Fixed Costs** — Costs which remain fixed over a certain range of volume, but then increase in a step fashion, at one or more trigger points.

**Semiliquid** — A material that is intermediate in physical properties, especially in flow properties, between liquids and solids.

**Semisolid** — Any substance having the attributes of both a solid and a liquid. Similar to semiliquid but being more closely related to a solid than a liquid.

**Senior Management** — A loosely defined term for the chief executive officer and his or her principal subordinates.

**Sensei** — A Japanese term for a personal teacher with mastery of a body of knowledge.

**Sensitivity** — The measure of a sensor's ability to detect small signals. Limited by the signal-to-noise ratio.

**Sensitivity Analysis** — The assessment of the impact that a change will have on the expected outcome of a process or project. The change may involve modifying the value of one or more of the variables or assumptions used to predict the outcome.

**Sensor** — Any device for measuring a variable and converting it into a signal that has a fixed relationship to the variable.

**Sequence Control** — The control of a series of machine movements, with the completion of one movement initiating the next. The extent of movements is typically not specified by numerical input data.

**Serial Port** — An input/output location (channel) that sends and receives data to and from a computer's central processing unit or a communications device, one bit at a time.

**Series** — A number of objects or events arranged or coming one after the other in succession.

**Server** — On a local area network (LAN), a computer running administrative software that controls access to the network and its resources, such as printers and disk drives, and provides resources to computers functioning as workstations on the network. On the Internet or other network, a computer or program that responds to commands from a client.

| | |
|---|---|
| **Service** | To make fit for use. To adjust, repair, or maintain. The performance of duties or provision of space and equipment helpful to production (or others). |
| **Service Contract** | A contract that directly engages the time and effort of a contractor to perform an identified task rather than furnish a physical product. |
| **Service History** | A record of measurements and findings that result from preventive or corrective maintenance actions. |
| **Service Level** | A measurement of the performance of a system. Based on defined goals, the service level gives the percentage to which they are achieved. |
| **Service Level – Inventory** | A measure of the performance of inventory systems. It is usually measured by the percentage of the inventory that is issued on demand. |
| **Service Level Agreement** | A formal agreement between an internal provider and an internal receiver (customer). |
| **Service Stock** | Commonly used parts and maintenance supplies kept nearby in high maintenance areas or outside the storeroom. Usually withdrawal of this stock does not require a requisition or paperwork. |
| **Serviceability** | A design characteristic that allows the easy and efficient performance of service activities. Service activities include those activities required to keep equipment in operation, such as lubrication, fueling, oiling, and cleaning. |
| **Servicing** | The performance of any act needed to keep an item in operating condition (lubricating, fueling, oiling, and cleaning) but not including preventive maintenance of parts or corrective maintenance tasks. |
| **Servomechanism** | A low power device that adjusts or controls a more powerful device in response to the changes detected in one or more variables. |
| **Servomotor** | A power-driven mechanism that supplements a primary control operated by a comparatively feeble force (as in a servomechanism). |

**Set Point** — The value of the process variable that is to be maintained by control action. The input variable used to control the value of a variable.

**Set-Up and Adjustment** — A process of changing from one manufacturing configuration to another to accommodate a change in product being produced on the same asset.

**Set-Up Time** — The time to set up the machine for product change over.

**Seven Tools of Quality** — Tools that help organizations understand their processes to improve them. The tools are:
- Cause and Effect Diagram
- Check Sheet
- Control Chart
- Flowchart
- Histogram
- Pareto Chart
- Scatter Diagram

**Severity** — The degree of the consequence of a potential loss or hazard. It is a numerical, subjective estimate of how severe the customer or end user will perceive the effect of a failure. SAE 1739 suggests using a 1-10 level of severity.

**Shakedown** — A form of a "run-in" or "return to service" check to ensure the equipment is capable of operating in desired condition.

**Shared Leadership Style** — Management approach in which the project manager holds that leadership consists of many functions and that these can be shared among the organization or team members. Some common functions are:
- Time keeping
- Record keeping
- Planning
- Scheduling
- Facilitating

Synonymous with *shared management style*.

**Shear – Mechanical**  The strain in, or failure of, a structural member at a point where the lines of force and resistance are perpendicular to the member.

**Shear Rate – Fluid**  The rate at which adjacent layers of fluid move with respect to each other, usually expressed as reciprocal seconds.

**Shear Stress – Fluid**  The frictional force overcome by sliding one layer of fluid along another.

**Shear Stress – Mechanical**  A stress which is applied parallel or tangential to a face of a material, as opposed to a normal stress which is applied perpendicularly.

**Shelf Life**  That period of time during which materials in storage remain in an acceptable condition.

**Shewhart Cycle**  A four-step process for quality improvement.
1. Plan – a plan to effect improvement is developed
2. Do – the plan is carried out, preferably on a small scale
3. Check – the effects of the plan are observed
4. Act – the results are studied to determine what was learned and what can be predicted

Synonymous with *Plan-Do-Check-Act (PDCA) cycle*.

**Shift**  A term applied to a work period where two or more workers are employed at different hours during the operating time of an organization.

**Shock Machine**  A device for subjecting a system to controlled and reproducible mechanical shock pulses. Synonymous with *shock test machine*.

**Shock Pulse**  An event that transmits kinetic energy into a system in a relatively short interval compared with the system's greatest natural period. A natural decay of oscillatory motion follows. The event is usually displayed as a time history, as on an oscilloscope.

**Shock Pulse Analysis (SPA)**  A technique for measuring shock pulses in rolling element bearings. Evaluation of the changes in the shock pulses provides data which is used to assess the condition of the bearing, especially to determine the condition of the lubrication film thickness between the rolling elements and the raceways.

| | |
|---|---|
| **Shock Response Spectrum** | The maximum response, in shock tests, of a series of single degree of freedom systems of the same damping to a given transient signal. |
| **Shock Test** | Tests performed to verify that a structure or a device can support transient vibrations encountered during its life in real environmental conditions. |
| **Shop Clean Up Time** | The assigned time to make an area ready for the next shift and cleaning of assigned spaces. |
| **Shop Drawings** | Drawings, diagrams, illustrations, schedules, performance charts, brochures, and other data prepared by the contractor, designer, manufacturer, supplier, or distributor that illustrate how specific portions of the work shall be fabricated and/or installed. |
| **Shop Floor Data Collection** | Automated collection of data on factory-production activities, including units produced, labor hours per unit or customer order, time and date of specific production activities, and maintenance and quality data. |
| **Shop Rules** | Regulations established by an employer dealing with day-to-day conduct in plant operations, safety, hygiene, records, etc. Synonymous with *work rules*. |
| **Short Circuit** | A fault or defect that causes part of the normal electrical circuit to be bypassed. |
| **Shoulder** | The side of a ball race, also a surface in a bearing application or shaft which axially positions a bearing and takes the thrust load. |
| **Shutdown** | A term designating a complete stoppage of production in a plant, system, or sub-system to enable planned or unplanned maintenance work to be carried out. Planned shutdowns are usually periods of significant inspection and maintenance activity, carried out periodically. |
| **Shutdown Maintenance** | Maintenance which can only be carried out when an asset is out of service. |
| **Sifting** | Screening through unnecessary materials and simplifying the work environment. Sifting is the separating of the essential from the nonessential. Synonymous with *sorting*. |

| | |
|---|---|
| **Sigma** | A term taken from the Greek letter σ (sigma), used in statistics as a measure of variation. The sigma value for a business process refers to the standard deviation and indicates how well that process is performing. |
| **Signal** | An indication (visual, audible, electrical) used to convey information. |
| **Signal Conditioner** | An amplifier following a sensor which prepares the signal for succeeding amplifiers, transmitters, readout instruments, etc. |
| **Signal Processing** | The acquisition, storage, analysis, alteration and output of digital data. In infrared and thermal testing, it is manipulation of temperature signal or image data for purposes of enhancing or controlling a process. |
| **Signal-to-Noise Ratio (SNR)** | The ratio of the amplitude of a desired signal at any point to the amplitude of noise signals at that same point. |
| **Signature – Electrical** | The characteristic spectrum of an alternating current. Motor defects can produce resonance and spikes in the modulation of its waveform which creates a unique current signature. |
| **Signature – Vibration** | A term usually applied to the vibration frequency spectrum which is distinctive and special to an equipment, component or system at a specific point of time, under specific operating conditions. |
| **Silicates** | Compounds made up of silicon, oxygen, and one or more metals with or without hydrogen. |
| **Silicon Controlled Rectifier (SCR)** | A semiconductor device that provides stepless power control of an electric power circuit. With this device, small electrical currents can be utilized to control high electrical loads. |
| **Silos – Organization** | A traditional functional form of departmental management in which the organization is managed vertically, often having functional specialists, and not employing an integrative, process-focused approach to product delivery. Silo organizations often have communication barriers. |
| **Silt** | Contaminant particles 5 micron (μm) or less in size. |

| | |
|---|---|
| **Silting** | A failure generally associated with a valve in which movements are restricted due to small particles that have wedged in between critical clearances (e.g., the spool and bore). |
| **Simple Harmonic Motion** | A periodic motion in which the restoring force is proportional to the displacement. It is the motion of a simple harmonic oscillator, a periodic motion that is neither driven nor damped. |
| **Simplification** | Principle of reducing the complexity of equipment or systems. |
| **Simulation** | A device, system, or computer program that represents certain features of the behavior of a physical or abstract system. |
| **Single Minute Exchange of Dies (SMED)** | The theory and techniques for performing setup operations in a minute. |
| **Single Point Failure** | The failure of an item which would result in the failure of the system and is not compensated for by redundancy or alternative operational procedure. |
| **Single Source** | One supplier from whom the entire quantity of goods or services is procured, even though other competitive suppliers are available. |
| **Single-Pass Test** | Filter performance tests in which contaminant which passes through a test filter is not allowed to recirculate back to the test filter. |
| **Single-Piece Flow** | A process in which products proceed, one complete product at a time, through various operations in design, order taking and production, without interruptions, backflows or scrap. |
| **SIPOC Diagram** | A SIPOC diagram, related to the process map, is a tool for defining business processes by listing:<br>• Suppliers<br>• Inputs<br>• Process<br>• Outputs<br>• Customers |

| | |
|---|---|
| **Situation Analysis** | A review process used to identify and define facts and variables that might influence a situation. |
| **Six Losses** | The major losses, in Total Productive Maintenance (TPM), that occur due to inadequate equipment operation or condition:<br>1. Breakdown<br>2. Setup and Adjustment<br>3. Minor Stoppages<br>4. Speed Reductions<br>5. Quality Defects and Rework<br>6. Yield Reductions |
| **Six Sigma** | A methodology that provides organizations with the tools to improve the capability of their business processes. |
| **Six Sigma Belts** | A system for designating experience levels of Six Sigma practitioners. Based on the perceived similarities of martial artists and Six Sigma experts, who attack problems through rigorous application of statistical techniques and the selective use of basic quality tools. See also *Green Belt*, *Black Belt*, and *Master Black Belt*. |
| **Six Sigma Culture** | An organization-wide culture that focuses on metrics and measurement, customer requirements, openness about defects, and empowerment and teamwork as it strives for continuous improvement by using the Six Sigma methodology. |
| **Six Sigma Quality** | A term generally used to indicate a process is well controlled (±6 sigma from the centerline in a control chart). The term is usually associated with Motorola, which named one of its key operational initiatives, Six Sigma quality. |
| **Skew** | The shape of a curve of an asymmetric distribution. Forms include positive skewness for skew right and negative skewness for skew left. |
| **Skew Left** | A distribution with negative skewness. It exhibits greater scatter and has a longer tail to the left of the mean. |
| **Skew Right** | A distribution with positive skewness. It exhibits greater scatter and a has a longer tail to the right of the mean. |

**Skill** — The knowledge and proficiencies required in the accomplishment of engineering, scientific, or any specific task.

**Skilled Labor** — Employees who have mastered one of the traditional crafts, usually through an apprenticeship, and who possess a thorough comprehensive knowledge of the job, have the ability to exercise considerable independent judgment, and the capability of assuming responsibility.

**Skills Inventory** — A planning tool that indicates people currently employed by the organization and classifies them according to their skills, job assignments, age, sex, and other factors relevant to human resource planning. Synonymous with *manning table*.

**Sleeve Bearing** — A journal bearing, usually a full journal bearing.

**Slew Rate** — The maximum rate at which an instrument's output can change by some stated amount.

**Slot Wedge** — The insulation that holds conductors firmly in the slots of an electric motor.

**Sludge** — Insoluble material formed as a result of deterioration reactions in an oil, contamination of an oil, or both.

**Smoke** — An air suspension (aerosol) of particles, often originating from combustion or sublimation. Carbon or soot particles less than $0.1\mu m$ in size result from the incomplete combustion of carbonaceous material such as coal or oil.

**SMRP Body of Knowledge** — The body of knowledge (BoK) for maintenance and reliability as identified by the Society for Maintenance & Reliability Professionals (SMRP). The SMRP BoK is structured into five pillars:

1. Business and Management
2. Manufacturing Process Reliability
3. Equipment Reliability
4. Organization and Leadership
5. Work Management

| | |
|---|---|
| **Snap Ring** | A removable ring used to axially position a bearing or outer ring in a housing. Also used as a means of fastening a shield or seal in a bearing. |
| **Sneak Circuit Analysis** | An analytical procedure for identifying latent paths that cause occurrence of unwanted functions or inhibit desired functions, assuming all components are operating properly. |
| **Sneak Paths** | A design error that permits the flow of current over an unintended path. |
| **Sneak Timing** | The occurrence of a circuit function at an improper time. |
| **Social Responsibility** | The principle that organizations should contribute to the welfare of society and not be solely devoted to maximizing profits. |
| **Socio-Cultural Barrier** | A real or perceived inhibitor of communication, association, or equality among groups of people. |
| **Soft Data** | Data that cannot be measured or specifically quantified, such as survey data that ask respondents to provide their "opinion" about something. |
| **Soft Failure** | A failure that occurs when a product under test ceases to operate correctly, but resumes correct operation when the stressing environment is eased. |
| **Software** | Computer programs. Instructions that make hardware work. Two main types of software are:<br>1. System software (operating systems) – which controls the workings of the computer.<br>2. Applications – word processing programs, spreadsheets, and databases, which perform the tasks for which people use computers. |
| **Software Quality Assurance (SQA)** | A planned and systematic approach to the evaluation of the quality of, and adherence to, software product standards, processes and procedures. SQA includes the process of assuring that standards and procedures are established and are followed throughout the software acquisition life cycle. |
| **Software Reliability** | The probability of failure-free software operation for a specified period of time in a specified environment. |

| | |
|---|---|
| **Sole Source** | The only source known to be able to perform a contract or the one source that, for a justifiable reason, is considered to be the most advantageous for the purpose of contract award. Synonymous with *sole-source supplier*. |
| **Sole Sourcing** | A deliberate decision to use only one vendor to supply material. |
| **Solenoid** | An electromechanical device made up of a coil which produces a magnetic field when electric current is passed through it. |
| **Solenoid Valve** | A shut-off valve whose position is controlled by a solenoid. Fluid flows through a normally open valve when the solenoid current is off. Fluid flows through a normally closed valve when the solenoid current is on. |
| **Solicitation** | Obtaining quotations, bids, offers, or proposals as appropriate. |
| **Solid** | Any substance having a definite shape which it does not readily relinquish. More generally, any substance in which the force required to produce a deformation depends upon the magnitude of the deformation rather than upon the rate of deformation. |
| **Solid Modeling** | Three-dimensional modeling in which solid characteristics of an object are built into the database so that complex internal structures can be realistically represented. |
| **Solid Waste** | Non-liquid, non-soluble materials ranging from municipal garbage to industrial wastes that contain complex and sometimes hazardous substances. Solid wastes also include sewage sludge, agricultural refuse, demolition wastes, and mining residues. |
| **Solid-State Sensor** | A sensor which has no moving parts. |
| **Solvency** | The ability of a fluid to dissolve inorganic materials and polymers, which is a function of aromaticity. |
| **Solvent** | A substance, usually a liquid, capable of dissolving or dispersing another liquid, gas, or solid to form a homogenous mixture. |

**Sone** — A unit of loudness.

**Sorbent** — A material which removes gases and vapors from air passed through a canister or cartridge.

**Sorting** — Sorting involves organizing essential materials. It allows the operator to find materials when needed because they are in the correct location. One of the 5S's used for workplace organization.

**Sound** — An oscillation in pressure, capable of evoking the sensation of hearing.

**Sound Intensity** — In a specified direction, the average rate of sound energy flow through a unit area perpendicular to that direction.

**Sound Level** — The quantity, in decibels, measured by a standardized sound level meter.

**Sound Monitoring** — The use of instruments, or the human ear, to detect changes in loudness, pitch, tone or frequency that could indicate pending problems with the functioning of equipment.

**Source Code** — Human-readable program statements written by a programmer or developer in a high-level or assembly language that are not directly readable by a computer. Source code needs to be compiled into object code before it can be executed by a computer.

**Source Follower** — A device for converting a high impedance electrical signal to low impedance. Synonymous with *impedance converter*.

**Source Impedance** — The combination of resistance and reactance that a source presents to the input terminals of a measuring instrument.

**Spall** — A flake or chip of metal which has been removed from one of the races of a rolling-element bearing.

**Span** — The difference between the maximum and minimum calibrated measurement or range values. An instrument having a calibrated range of 100 to 300 has a range of 200.

**Span of Control** — The number of subordinates that a given person supervises.

**Spare Part** — Replacement items identified in a CMMS or on a bill of material.

**Spatial Resolution** — A measure of the ability of an infrared system to see detail, usually specified by its instantaneous field of view.

**SPC & Control Charts** — A method of monitoring, controlling and, ideally, improving a process through statistical analysis using data from statistical process control (SPC) and control charts. Its four basic steps include:

1. Measuring the process.
2. Eliminating variances in the process to make it consistent.
3. Monitoring the process.
4. Improving the process to its best target value.

**Special Causes** — Causes of variation that arise because of special circumstances. They are not an inherent part of a process. Synonymous with *assignable causes*.

**Specific Gravity** — The ratio of the weight of a given volume of material to the weight of an equal volume of water.

**Specific Heat** — The amount of heat required to raise a unit mass of a given substance by a unit temperature.

**Specific Learning Objective** — The detailed knowledge, skill, or attitude necessary to perform a job.

**Specification** — A description of the technical requirements for a material, product, or service, including the criteria for determining that the requirements have been met.

**Spectral Map** — A three-dimensional plot of amplitude (Z axis) vs. time (or shaft speed) (Y axis) vs. frequency (X axis).

**Spectrographic Analysis** — A method of identifying elements and compounds and determining their atomic and molecular structure by measuring the radiant energy absorbed or emitted by them at characteristic wavelengths of the electromagnetic spectrum.

**Spectrographic Oil Analysis Program** — The procedures for extracting fluid samples from operating systems and analyzing them spectrographically for the presence of key elements.

| | |
|---|---|
| **Spectrophotometry** | The measurement of electromagnetic radiant energy as a function of wavelength, particularly in the ultraviolet, visible and infrared wavelength. It is used to determine color, turbidity, opacity, and other characteristics of fluids. |
| **Spectrum Analysis** | The measurement of the amplitude of the components of a complex waveform throughout the frequency range of the waveform. It is the most commonly used analysis method for machinery diagnostics, such as to identify the majority of rotating equipment failures before they fail. |
| **Spectrum Analyzer** | An instrument which displays the frequency spectrum of an input signal, usually amplitude vs. frequency. |
| **Specular Reflector** | A surface that reflects radiation at an angle equal to the angle of incidence, a mirror image. |
| **Spider Chart** | A two-dimensional chart of three or more quantitative variables represented on axes starting from the same point. Synonymous with *radar chart*. |
| **Spill Tanks** | Holding tanks used to collect chemicals discharged from a chemical storage facility. Usually located away from a facility on a tank farm, spill tanks are connected to different waste line systems to prevent the mixing of incompatible chemicals. |
| **Spindle Oil** | A light-bodied oil used principally for lubricating textile spindles and light, high-speed machinery. |
| **Spin-On Filter** | A throw-away type bowl and element filter assembly that mates with a permanently installed head. |
| **Splash Lubrication** | A system of lubrication in which parts of a mechanism dip into and splash the lubricant onto themselves and/or other parts of the mechanism. |
| **Spline** | One of a series of ridges on a driveshaft which mesh with and equalize the rotation speed of a mating piece, thereby transferring torque. Synonymous with *rotating spline*. |
| **Split Shift** | The daily working time that is not continuous, but split into two or more working periods. |
| **Sponsor** | The person who supports a team's plans, activities and outcomes. The team's backer. |

**Spread**  A measure of variability and dispersion equal to the absolute difference between the largest observed value and the smallest observed value in a given sample.

**SQL**  An acronym for *Structured Query Language*, a database sublanguage used in querying, updating, and managing relational databases. The de facto standard for database products.

**Squeak**  A sound resembling that of an un-lubricated hinge, or as made by PVC or other material rubbing on glass.

**Stability**  A measure of the change or wander of important parameters of the parts in the lot, or attribute(s) for a product or process over a period of time.

**Stable Operations**  Processes from which special-cause variation has been removed and common-cause variation has been sufficiently reduced to make outputs consistent and predictable.

**Staff**  An organizational unit that serves in an auxiliary and facilitative role in relation to line and operating units.

**Stages of Team Growth**  Four stages that teams move through as they develop maturity over time:
1. Forming
2. Storming
3. Norming
4. Performing

**Stakeholder**  Any individual, group, or organization that will have a significant impact on, or will be significantly impacted by, the quality of the product or service an organization provides.

**Standard**  A document approved by a recognized body that provides for common and repeated use, rules, guidelines, or characteristics for products, processes, or services. Compliance may not be mandatory.

**Standard – Measurement**  A specification against which the outputs of a process are compared and declared acceptable or unacceptable.

**Standard Deviation**  A computed measure of variability indicating the spread of the data set around the mean.

| | |
|---|---|
| **Standard Industrial Classification (SIC)** | A coding system of the U.S. government used to identify specific economic sectors. Coding for most manufacturers encompasses the four-digit numbers of 2000 through 3999. Replaced by the North American Industry Classification System (NACIS) in 2001. |
| **Standard Job** | A preplanned maintenance job with all details required for work execution delineated and stored for repeated use. |
| **Standard Operating Conditions (SOC)** | The set of conditions (voltage, temperature, humidity, etc.) over which specified parameters maintain their stated performance rating. |
| **Standard Operating Procedures (SOP)** | Detailed step-by-step procedure for repetitive operations. |
| **Standard Procedure** | A documented prescription that a certain type of work be done in the same way wherever it is performed. |
| **Standard Units Produced** | The quantity produced as output. Examples would include gallons, liters, pounds, kilograms, or other standard units of measures. |
| **Standard Work** | A precise description of each work activity specifying cycle time, task time, the work sequence of specific tasks, and the minimum inventory of parts on hand needed to conduct the activity. |
| **Standard Work Instructions** | A tool that enables technicians to have a good understanding of how work tasks are to be performed. It ensures that the quality level is understood and serves as an excellent training aid. It enables absentee replacement individuals to easily adapt and perform the repair or an assembly operation. |
| **Standardization** | The practice of designing equipment or systems using common parts, components, or equipment to facilitate maintenance and reduce the number of maintenance and capital spares. |
| **Standards Body** | An organization that develops standards. |
| **Standby** | Assets that are used as backups to others. Those that are installed or available but not in use. |

**Standby Redundancy** — A system in which a redundant component starts operating only when the active component fails. Under standby redundancy, the redundant components do not share any of the load, and they start operating only when active components fail, are manually started or by an automatic control system.

**Standby Time** — The time for which an item or system is available if required, but not used.

**Standing Wave** — A wave that is characterized by lack of vibration at certain points (nodes), between which are areas of maximum vibration (antinodes). Standing waves are produced at certain forcing frequencies when the resulting resonant vibratory response is confined within boundaries, as in the vibrating string of a musical instrument. Synonymous with *stationary wave*.

**Standing Work Order** — A work order, opened for a specific period of time to capture labor and material costs for recurring or short duration maintenance work and work that is not associated with a specific piece of equipment, where tracking work history or formalizing individual work orders is not cost effective or practical. Examples include shop housekeeping, meetings, and training. Synonymous with *blanket work order*.

**Start Date** — The point in time associated with an activity's start.

**Startup** — The period, after the date of initial operation, during which a unit is brought up to acceptable production capacity and quality.

**Static Discharge Head** — The vertical distance from the centerline of the pump impeller to the point of discharge.

**Static Friction** — The force just sufficient to initiate relative motion between two bodies under load. The value of the static friction at the instant relative motion begins is termed break-away friction.

**Static Load** — A load that is at rest and exerts downward pressure only.

**Statistical Process Control (SPC)** — A method whereby data is collected and statistics are used to understand process variation.

| | |
|---|---|
| **Statistical Quality Control (SQC)** | A method of monitoring, controlling and, improving a process through statistical analysis. Its four basic steps include:<br>1. Measuring the process.<br>2. Eliminating variances in the process to make it consistent.<br>3. Monitoring the process.<br>4. Improving the process to its best target value. |
| **Statistics** | A field that involves the tabulating, depicting, and describing of data sets. A formalized body of techniques characteristically involving attempts to infer the properties of a large collection of data from inspection of a sample of the collection. |
| **Stator** | That part of an AC induction motor's magnetic structure which does not rotate. It usually contains the primary winding. The stator is made up of laminations, with a large hole in the center in which the rotor can turn. There are slots in the stator in which the windings for the coils are inserted. |
| **Status** | The condition of a project at a specified point in time, relative to time, cost, or performance. |
| **Status Report** | A description of the current status of a project. |
| **Steady-State Failure Rate** | The constant failure rate after the infant mortality period. |
| **Steady-State Vibration** | Periodic vibration for which the statistical measurement properties (such as the peak, average, RMS and mean values) are constant. |
| **Steam** | Water, in the vapor state. |
| **Steam Trap** | A device used to drain water from a steam system. |
| **Step Stress Test** | A form of accelerated test that reveals the uniformity and strength of an item, but does not normally yield failure rate. The test repeatedly employs increased stresses according to a prearranged plan. |

| | |
|---|---|
| **Step Stressing** | Increasing stresses in a series of preselected increments. |
| **Stepped Sine Testing** | The measurement of the frequency response functions of structures using stepped sine excitation. |
| **Stereolithography** | A form of rapid prototyping involving the building of solid plastic objects from a CAD file by projecting a laser-generated beam of ultraviolet radiation onto the surface of a vat of photosensitive resin. |
| **Stereotyping** | The tendency to attribute characteristics to an individual on the basis of an assessment of the group to which the individual belongs. |
| **Stiffness** | The resistance of an elastic body to deformation by an applied force. |
| **Stochastic** | Developed in accordance with a probabilistic manner or model. |
| **Stock** | A term used to describe parts that are normally kept on-hand in a storeroom. |
| **Stock Checkout** | The method used to process issues and returns for inventory items from and to a storeroom. |
| **Stock Code** | A number assigned in an inventory system to an item having a specific form, fit, and function. All items of that specific form, fit, and function have the same stock code. |
| **Stock Inventory Value** | The current book value of maintenance, repair, and operating (MRO) materials held in stock at the plant site (including consignment and vendor-managed stores). Synonymous with *stocked MRO inventory value* and *storeroom inventory value*. |
| **Stock Issue Card** | The authorized document for making stock material withdrawals. |
| **Stock Item** | An inventoried item that the storeroom manages at a specified quantity. |
| **Stock Keeping Unit (SKU)** | An inventory management term for individual stock items carried in inventory with assigned inventory numbers. |

| | |
|---|---|
| **Stock Level** | The inventory level designed to ensure that a storeroom has parts and supplies available as needed. |
| **Stock Out** | A measure of how many times the storeroom is out of stock for an item when that item is requested. |
| **Stockless Purchasing** | Buying material, parts, supplies, etc., for direct use by the departments involved, as opposed to receiving them into stores and subsequently issuing them to the departments. This reduces inventory investment. |
| **Stoke (St)** | A kinematic measurement of a fluid's resistance to flow. Defined by the ratio of the fluid's dynamic viscosity to its density. |
| **Stokes' Law** | A formula that predicts how fast an oil droplet will rise or settle through water based on the density and size of the oil droplet size and the distance the object must travel. |
| **Stop Work Order** | A formal request to stop work because of nonconformance, funding, or technical limitations. |
| **Storage Life** | The length of time an item can be stored under specified conditions and still meet specified requirements. Synonymous with *shelf-life*. |
| **Stored Procedure** | A precompiled collection of SQL statements and optional control-of-flow statements stored under a name and processed as a unit. They are stored in a SQL database and can be run with one call from an application. |
| **Storeroom Clerk** | An employee who has responsibilities for the day to day activities of the storeroom. Synonymous with *storekeeper, storeroom attendant*, and *storeroom associate*. |
| **Storeroom Inventory Accuracy** | The total number of storeroom SKU's counted with 100% accuracy (with respect to quantity and location) divided by the total number of storeroom SKU's counted. |
| **Storeroom Inventory Fill Rate** | The total number of storeroom issue requests (manual or on-line) 100% filled divided by the total number of storeroom issue requests. |
| **Storeroom Inventory Turnover** | The value of storeroom materials used divided by the average storeroom inventory. Usually calculated annually. |

| | |
|---|---|
| **Stores Inventory Turns** | Identifies how quickly specific types of inventory are flowing through the inventory system. Normally this is divided into at least two categories:<br>1. Operating supplies that are supposed to turn frequently.<br>2. Spare parts which usually have a lower turnover. |
| **Stores Investment** | The amount of capital invested in spares, strategic parts, and consumables used for maintenance. |
| **Stores Requisition** | The prime document raised by user departments authorizing the issue of specific materials, parts, supplies or equipment from the store or warehouse. |
| **Storyboarding** | A graphical summary of progress on a project. It is used to track data, decisions, and actions, and to create a pictorial record of an improvement project. |
| **STPD Conditions** | Standard temperature and pressure, dry (STPD) conditions, which are at standard temperature (0 °C), barometric pressure at sea level (101.3 kPa) and without water vapor (dry). |
| **Straight-Line Method Depreciation** | A depreciation method in which an equal amount of an asset's cost is considered an expense for each year of its useful life. |
| **Strain** | The physical deformation, deflection, or change in length, resulting from stress. |
| **Strainer** | A coarse filter element (pore size over approximately 40 μm). |
| **Strain-Gage Transducer** | A changing-resistance sensor whose signal depends upon sensitive element deformation. |

**Strategic Plan**

A document used by an organization to align its organization and budget structure with organizational priorities, missions, and objectives. A strategic plan should include a:

- Mission statement
- Long term goals and objectives – their description
- Strategies or means – how the organization plans to achieve their goals and objectives

The strategic plan should also identify external factors that could affect achievement of long-term goals.

**Strategic Planning**

The process by which an organization envisions its future and develops strategies, goals, objectives, and action plans to achieve that future.

**Strategy**

An action plan to set the direction for the coordinated use of resources through programs, projects, policies, procedures, and organizational design and establishment of performance standards.

**Stratification**

The process of classifying data into subgroups based on characteristics or categories.

**Streamlined RCM**

An optimized form of reliability centered maintenance (RCM) that begins with a risk rank prioritization to assure resources are applied most effectively to equipment and systems with the highest potential for value and return, and builds from templates and predetermined maintenance lists to assure all potential failures and corrective action are considered.

**Stress**

Force per unit area. When the force is one of compression, it is known as pressure.

**Stress Screening**

A tool for precipitating latent defects such as poorly-soldered connections in electronics production. The technique utilizes random vibration combined with rapid temperature ramping.

**Stress To Failure (STF) Test**

A destructive test where one or more stress factors, such as temperature or voltage, are increased until an entire sample has failed. The mean value of the stress to failure (STF), and the distribution of failure events about this mean, are used as measures of part reliability.

| | |
|---|---|
| **Stretch Goal** | A goal that is hard to reach, but not seen as impossible. Synonymous with *Big, Hairy Audacious Goal (BHAG)*. |
| **String-Based PM** | Preventive maintenance (PM) tasks that are strung together on several machines. Examples of string PM's would include lubrication, filter change, or vibration routes. |
| **Strip Chart** | A continuous chart on which a recording instrument places a permanent record, as opposed to a fixed-period circular chart. |
| **Structural Variation** | The variation caused by regular, systematic changes in output, such as seasonal patterns and long-term trends. |
| **Subcontractor** | A contractor, distributor, vendor, or firm that furnishes supplies or services to a prime contractor or another subcontractor. |
| **Subharmonic** | A sinusoidal quantity having a frequency that is an integral submultiple (1/2, 1/3, etc.) of a fundamental (1) frequency. |
| **Subject Matter Expert (SME)** | An individual widely recognized for knowledge and expertise in a particular area. |
| **Suboptimization** | A solution to a problem that is best from a narrow point of view but not from a higher or overall company point of view. |
| **Subsychronous** | Components of a vibration signal whose frequency is less than the shaft rotational speed. |
| **Sub-System** | A combination of equipment or components that performs an operational function within a system and is a major subdivision of the system. |
| **Success Indicator** | A measure of a parameter that can be trended to identify success in reaching a personal, team, department, or plant goal. Examples include routine maintenance expenses, schedule compliance, and maintenance backlog. Synonymous with *key performance indicator*. |
| **Succession Planning** | A key element of the workforce development process. It identifies and prepares suitable employees through mentoring, training, and job rotation to replace key players in the organization. |

|  |  |
|---|---|
| **Successor Activity** | An activity that starts after the start of a current activity. |
| **Suction Filter** | A pump intake-line filter in which the fluid is below atmospheric pressure. |
| **Sulfated Ash** | The ash content of fresh, compounded lubricating oil as determined by ASTM Method D 874. |
| **Sulfurized Oil** | Oil to which sulfur or sulfur compounds have been added. |
| **Sunk Cost** | A cost that once expended can never be recovered or salvaged. |
| **Superclean** | Having a concentration of fewer than ten particles per milliliter that are smaller than 10 micron in size. |
| **Superintendent** | A manager who is responsible for a group or department. |
| **Supervision** | The function of leading, coordinating, and directing the work of others to accomplish designated objectives. |
| **Supervisor** | A first-line leader who is responsible for work execution. |
| **Supervisor to Craft Ratio** | The number of maintenance workers a single supervisor is managing. |
| **Supplier** | A source of materials, service, or information input provided to a process. |
| **Supplier Partnerships** | Agreements with suppliers whereby operations are linked together, information is openly shared, problems and issues are commonly solved, and joint performance is mutually approved. They usually include multiyear purchase agreements. |
| **Supplier Quality Assurance** | The confidence a supplier's product or service will fulfill its customers' needs. This confidence is achieved by creating a relationship between the customer and supplier that ensures the product will be fit for use with minimal corrective action and inspection. |
| **Supply Chain** | The series of suppliers relating to a given process. |

**Supply Chain Logistics Systems**  A class of manufacturing software designed to optimize scheduling and other activities throughout the supply chain, or value chain, including transportation and distribution functions.

**Supply Chain Management (SCM)**  The management of a network of interconnected businesses involved in the ultimate provision of product and service packages required by end customers. SCM spans all movement and storage of raw materials, work-in-process inventory, and finished goods from point of origin to point of consumption.

**Supply Current**  The typical current that must be supplied to a sensor (along with the supply voltage).

**Supply Delay Time**  That element of delay time during which a needed replacement item is being obtained.

**Supply Support**  Management methods, practices, and procedures employed in determining requirements of goods and services and their acquisition, receipt, storage, issuance, and final disposal. Supply support is an important element of integrated logistics support.

**Support Equipment**  Items required to maintain systems in effective operating condition under various environments. Includes general and special-purpose vehicles, power units, stands, test equipment, tools, and test benches needed to facilitate or sustain maintenance action, detect or diagnose malfunctions, and monitor the operational status of equipment and systems.

**Surface Fatigue Wear**  The formation of surface or subsurface cracks and fatigue crack propagation. It results from cyclic loading of a surface.

**Surface Filtration**  Filtration which primarily retains contaminant on the influent surface.

**Surface Tension**  The contractile surface force of a liquid by which it tends to assume a spherical form and present the least possible surface. It is expressed in dynes/cm, or ergs/cm$^2$.

**Surface-Mount Technology**  A method of mounting elements on printed circuit boards.

| | |
|---|---|
| **Surfactant** | Surface-active agent that reduces interfacial tension of a liquid. A surfactant used in a petroleum oil may increase the oil's affinity for metals and other materials. |
| **Surge** | A momentary rise of pressure in a fluid circuit. |
| **Surge Testing** | The use of high-voltage DC pulses to detect grounds and winding faults in AC motors. Synonymous with *surge comparison testing*. |
| **Surveillance** | The continual monitoring of a process. A type of periodic assessment or audit conducted to determine whether a process continues to perform to a predetermined standard. |
| **Survey** | The act of examining a process or questioning a selected sample of individuals to obtain data about a process, product, or service. |
| **Surveying – Land** | The process of recording observations, making measurements, and marking the boundaries of tracts of land. |
| **Sustainability** | Development that meets the needs of the present without compromising the ability of future generations to meet their own needs. |
| **Sustainable Manufacturing** | The creation of manufactured products that use processes that are non-polluting, conserve energy and natural resources, and are economically sound and safe for employees, communities, and consumers. |
| **Sustaining** | The continuation of sifting, sweeping, sorting and sanitizing. It is the most important and the most difficult, because it addresses the need to perform the 5S's on an on-going and systematic basis. |
| **SWAG** | An estimate of time, or cost, of completing a project or element of work based solely on the experience of the estimator. Typically done in haste, SWAG estimates are usually no more accurate than order-of-magnitude estimates. An acronym for *scientific wild anatomical guess (SWAG)*. |
| **Swap** | A maintenance technique in which a known good, or high quality part, replacement assembly is substituted for a suspected bad assembly. Synonymous with *swap out*. |

**Swarf** — The fine cuttings and grindings that result from metal working operations.

**SWOT Analysis** — *S*trengths, *w*eakness, *o*pportunities and *t*hreats (SWOT) analysis used to determine where to apply special efforts to achieve desired outcomes. Entails listing:
- Strengths and how to take advantage of them.
- Weaknesses and how to minimize their impacts.
- Opportunities and how to capitalize on them.
- Threats and how to deal with them.

**Symptom** — An observable phenomenon arising from and accompanying a defect.

**Synchronous** — Vibration components (on rotating machinery) that are related to shaft speed.

**Synchronous Motor** — A motor which operates at a constant speed up to full load.

**Synchronous Sampling** — Control of a computer's rate of data sampling to achieve order tracking.

**Synergy** — A process in which more is accomplished by cooperation than could be done by separate efforts.

**Synthetic Hydrocarbon** — An oil molecule with superior oxidation quality tailored primarily out of paraffinic materials.

**Synthetic Lubricant** — A lubricant produced by chemical synthesis, rather than by extraction or refinement of petroleum, to produce a compound with planned and predictable properties.

**System** — A group of interdependent processes and/or people that is intended to perform a common function.

**System Architect** — The role of managing the capability sustainment of major assets. The system architect is responsible for providing oversight of requirements, assessments, and alternate solutions to facility needs in support of investment planning, operations, and maintenance. The technical guru of the facility.

| | |
|---|---|
| **System Architecture – Computer** | The manner in which hardware or software is structured. How the system or program is constructed, how its components fit together, and what protocols and interfaces are used for communication and cooperation among the components, including human interaction. |
| **System Balancing** | Assigning numbers of operators or machines to each operation of an assembly line so as to meet the required production rate with a minimum of idle time. |
| **System Design** | The translation of customer requirements into comprehensive, detailed, functional performance or design specifications, which are then used to construct the specific solution. |
| **System Effect** | The consequence(s) a failure mode has on the highest indenture level of a system. |
| **System Effectiveness** | The probability that a system will meet its operational demand within a given time under specified operating conditions. |
| **System Engineering** | The logical sequence of activities and decisions to transform an operational need into a description of system performance parameters and a preferred system configuration. |
| **System Kaizen** | Improvement aimed at an entire value stream. |
| **System Manager** | The person assigned responsibility for the technical and configuration management of an asset or grouping of similar assets. He/she is recognized as the technical expert for the asset or assets to which they are assigned. |
| **System of Profound Knowledge** | A system, defined by W. Edwards Deming, which consists of an appreciation for systems, knowledge of variation, theory of knowledge and understanding of psychology. |
| **System Pressure** | The pressure which overcomes the total resistances in a system. It includes all losses as well as useful work. |
| **System Requirements Document** | The document containing verifiable, functional and performance requirements and design constraints that a system must have within a defined environment, or set of conditions, in order to provide a needed operational capability and comply with applicable standards. |

| | |
|---|---|
| **System Safety Analysis** | A systematic process of identifying system hazards with the goal of mitigating or eliminating them. |
| **System Safety Engineering** | The application of scientific and engineering principles during the design, development, manufacture, and operation of a system to meet or exceed established safety goals. |
| **System Verification Review** | A functional configuration audit, conducted at the system level, to verify the effective assembly and integration of the configuration item and related computer software into a functioning system. |
| **Systems Analysis** | A method of problem-solving that encompasses the identification, study, and evaluation of interdependent parts and their attributes, functioning as an ongoing process, and constituting an organic whole. |
| **Systems Approach** | A wide-ranging, synthesizing method of addressing problems that considers multiple and interacting relationships. Commonly contrasted with the analytic approach. |
| **Systems Integration** | The ability of computers, instrumentation, and equipment to share data or applications with other components in the same or other functional areas. |
| **Systems Management** | Coordination and maintenance of all the software on a client/server network, including performance monitoring, applications distribution, version control, user administration, job scheduling, data back-up, security, and configuration management. |
| **Systems Thinking** | A method that emphasizes the value of viewing a system as a whole before examining its parts. By doing so, the environmental context of the system is better understood, resulting in greater appreciation and understanding of how the individual parts interact with the whole. |

# T

**Tacit Knowledge**  From the Latin *tacitare*, which refers to something that is very difficult to articulate, or put into words or images. Typically, highly internalized knowledge, such as knowing how to do something or recognizing analogous situations.

**Tactics**  The choices made to implement a strategy and manage the people, processes, and physical asset infrastructure that make up your business.

**Tactile**  Of, or relating to, the sense of touch.

**Tag – Equipment**  An equipment or device identification number. Synonymous with *brass tag*.

**Taguchi Loss Function**  The parabolic approximation of the money lost to the customer due to a quality characteristic deviating from its target value. This function shows no loss at the target value, but loss increases exponentially as deviation increases from the target value, even within tolerance limits.

**Taguchi Methods**  The American Supplier Institute's trademarked term for the quality engineering methodology developed by Genichi Taguchi. In this engineering approach to quality control, Taguchi calls for off-line quality control, on-line quality control, and a system of experimental design to improve quality and reduce costs.

**Tailoring**  The making or adapting of an item or process to suit a particular purpose.

**Takt Time**  The rate of customer demand. Takt is the heartbeat of a lean system. Takt time is calculated by dividing production time by the quantity the customer requires in that time.

**Tampering**  The action taken to compensate for variation within the control limits of a stable system. Tampering increases rather than decreases variation, as evidenced in the funnel experiment.

| | |
|---|---|
| **Target Date** | The date an activity is planned to start or end. |
| **Task** | A specific, definable activity to perform an assigned piece of work, often finished within a certain time. |
| **Task Analysis** | An analytical process employed to determine the specific behaviors required of human components in a human-machine system. |
| **Task Element** | A subset of a task, the work to be accomplished in a small unit. |
| **Task Force** | A team of skilled contributors who are charged with investigating a problem for the specific purpose of developing and implementing a solution. |
| **Task List** | Directions to an executor of a task (e.g., maintainer) telling him or her what to do and in what sequence (e.g., check oil level, clean, adjust, lubricate, replace, etc.). |
| **Task Type** | Identification of a task by resource requirement, responsibility, discipline, jurisdiction, function, or any other characteristic used to categorize it. |
| **Taxonomy** | A basic classification system that enables the conceptual identification of hierarchies and dependencies. |
| **TCP/IP** | An acronym for *t*erminal *c*ontrol *p*rotocol/*i*nternet *p*rotocol, a communications standard that is used by the Internet. A protocol is the understanding that computers have for how information will be delivered over the communications network, which enables computers with different operating systems to communicate with each other and eliminate errors in data transmission. |
| **Team** | A group of individuals organized to work together to accomplish a specific objective. |
| **Team Building** | A planned and deliberate process of encouraging effective working relationships while diminishing difficulties or roadblocks that interfere with the team's competence and resourcefulness. |
| **Team Development** | The development of individual and group skills to improve team performance. |

| | |
|---|---|
| **Teardown** | The act or process of taking apart a piece of equipment or structure. |
| **Technical Data Package** | Documents that provide a technical description of an item (product and process) adequate for supporting an acquisition strategy, production, engineering, and logistics support. It consists of all applicable technical data such as drawings, procedures, and manuals. |
| **Technical Requirements** | Description of the features of the deliverable in detailed technical terms. |
| **Technique** | A practical method, or art, applied to some particular task. |
| **Technique for Human Error Rate Prediction (THERP)** | A technique used in the field of *h*uman *r*eliability *a*ssessment (HRA), for the purposes of evaluating the probability of a human error occurring throughout the completion of a specific task. From such analyses, measures can then be taken to reduce the likelihood of errors occurring within a system and therefore lead to an improvement in the overall levels of safety. |
| **Temperature** | An arbitrary measurement of the amount of molecular energy of a body or the degree of heat possessed by it. A measurement of heat and cold. |
| **Temperature Monitoring** | A technique of looking for potential failures that cause a rise in the temperature of the equipment or device. |
| **Temperature Range** | The temperatures between which a piece of equipment or device will operate accurately. |
| **Templates** | Sets of guidelines that provide sample outlines, forms, checklists, and other documents. |
| **Temporary** | A finite period of time, with a defined beginning and end. |
| **Temporary Repair** | Repair activities that temporarily address breakdowns and repair ailing equipment. The purpose is to get equipment back into operation until routine or corrective maintenance can be performed to effect permanent repair. Synonymous with *band-aiding*. |

| | |
|---|---|
| **Tensile Strength** | Resistance of a material to a force that tends to pull it apart. Usually expressed as the measure of the largest force that can be applied in this way before the material breaks apart. |
| **Terminal – Computer** | An input device, in networking, consisting of a video adapter, a monitor, and a keyboard. The adapter and monitor (and keyboard) are typically combined in a single unit. A terminal does little or no computer processing, instead, it is connected to a computer with a communications link over a cable. Terminals are used primarily in multiuser systems. |
| **Tesla** | A unit of measure of magnetic field strength. |
| **Test for Normal Probability** | A statistical test used to check whether observations follow a normal distribution. |
| **Test Interval** | The elapsed time between tests on an item to evaluate its state or condition. It is the inverse of test frequency. |
| **Test Measurement and Diagnostic Equipment (TMDE)** | Those devices used to maintain, evaluate, measure, calibrate, test, inspect, diagnose, or otherwise examine materials, supplies, equipment, and systems to identify or isolate actual or potential malfunction, or decide if they meet operational specification. Examples include strain gauges, thermometers, and measuring devices. |
| **Testability** | A characteristic of an item's design which allows the status (operable, inoperable, or degraded) of that item to be confidently determined in a timely manner. |
| **Theoretical Capacity** | The volume of activity or throughput that could be attained under ideal operating conditions, with minimum allowance for inefficiency. It is the largest volume of output possible. |
| **Theoretical Discharge Head** | The maximum height to which water can be lifted inside a tube under perfect conditions (perfect vacuum) at sea level. |

**Theory Of Constraints (TOC)**  A method for identifying and overcoming key bottlenecks and constraints which inhibit an asset, process, or organization from achieving its goal. Originally developed by Dr. Eliyahu Goldratt, and published in his book, *The Goal*. Based on the theory that a system has a single goal, and that systems are composed of multiple linked activities, one of which acts as a constraint on the whole system.

**Theory X Management**  An approach to managing people described by MacGregor. Based on the philosophy that people dislike work, will avoid it if they can, and are interested only in monetary gain from their labor. Accordingly, the Theory X manager will act in an authoritarian manner, directing each activity of his or her staff.

**Theory Y Management**  An approach to managing people described by MacGregor. Based on the philosophy that people will work best when they are properly rewarded and motivated, and that work is as natural as play or rest. Accordingly, the Theory Y manager will act in a generally supportive and understanding manner, providing encouragement and psychic rewards to his or her staff.

**Theory Z Management**  An approach to managing people described by Arthur and Ouehi. Based on the philosophy that people need goals and objectives, motivation, standards, the right to make mistakes, and the right to participate in goal setting. More specifically, it describes a Japanese system of management characterized by the employee's heavy involvement in management, which has been shown to result in higher productivity levels when compared to U.S. or western counterparts. Successful implementation requires a comprehensive system of organizational and sociological rewards. Its developers assert that it can be used in any situation with equal success. Synonymous with *participative management style*.

**Thermal Capacitance**  The ability of material to store thermal energy. Defined as the amount of heat required to raise the temperature of one cubic feet of material one degree Fahrenheit. It is the product of a material's specific heat multiplied by density. Synonymous with *volumetric heat capacity*.

| | |
|---|---|
| **Thermal Conductivity** | A measure of the ability of a material to transfer heat. Defined as the rate at which heat flows though a material of unit area and thickness, with a temperature gradient, over a unit of time. |
| **Thermal Cycling** | The subjecting of a product or process to predetermined temperature changes, between hot and cold extremes. |
| **Thermal Expansion** | The expansion caused by the increase in temperature. This may be linear or volumetric. |
| **Thermal Resistance** | An index of a material's resistance to heat flow. The reciprocal of thermal conductivity. |
| **Thermal Shock** | The stress-producing phenomenon resulting from a sudden, large temperature change. |
| **Thermal Stability** | The ability of a fuel or lubricant to resist oxidation under high temperature operating conditions. |
| **Thermistor** | A thermally sensitive, resistor semi-conductor used for temperature measurement. |
| **Thermocouple** | The junction of two wires of dissimilar metals that develops an electrical potential that is a function of the temperature. |
| **Thermocouple – Instrumentation** | An instrument for measuring temperature by means of the electrical potential produced at a heated junction of two dissimilar metals. |
| **Thermography** | The use of infrared cameras to measure equipment temperatures remotely and without contact. The infrared energy radiating from the surface of the target is measured and converted to an equivalent surface temperature. |
| **Thermometer** | A device that measures temperature, or temperature gradient, using a variety of different principles and media. |
| **Thermostat** | An apparatus for maintaining, and keeping constant, any practicable temperature. |
| **Thermowell** | A cavity within a vessel or line, but sealed off from it, for the purpose of inserting a thermocouple or thermometer for temperature measurement. |

**Thin Film Lubrication**  A condition of lubrication in which the film thickness of the lubricant is such that the friction between the surfaces is determined by the properties of the surfaces as well as by the viscosity of the lubricant.

**Thixotropy**  The tendency of grease or other material to soften or flow when subjected to shearing.

**Threat**  A thing, or condition, that has the potential to cause harm or result in an adverse outcome or loss.

**Three Phase – Power**  A common method of transmission of alternating-current (AC) electricity. A type of polyphase system, it is the most common method used by electric power distribution grids to distribute power. It is also used to power large motors and other equipment with large loads demands.

**Three-Body Abrasion**  A particulate wear process by which particles are pressed between two sliding surfaces.

**Threshold**  The smallest change in a measured variable that gives a measurable change in output signal.

**Throttling Control**  A type of control that can position the final control element at any position between maximum and minimum limits.

**Throughput**  The rate at which work proceeds through a manufacturing process or machine.

**Throughput – Computer**  A measure of the data processing rate in a computer system.

**Throughput – Network**  The data transfer rate of a network, measured as the number of bits per second transmitted.

**Thrust Bearing**  An axial-load bearing.

**Thrust Position**  Location in direction of a shaft centerline.

**Tied Activity**  An activity that must start within a specified time, or immediately after its predecessor's completion. Used in PERT charts.

| | |
|---|---|
| **TIG Welding** | An arc welding process that uses a non-consumable tungsten electrode to produce the weld and an inert gas as a shielding gas. |
| **Time** | The universal measure of duration. |
| **Time Available to Schedule** | The number of craft hours available to schedule during a period of time. Do not include vacation, sickness, or time off due to injuries. |
| **Time Card** | A card for recording actual employee hours of work. Many time cards are designed to be inserted into a time-recording device, which stamps the current time on the card, indicating check-in and check-out times. Can also be in the form of a time sheet. |
| **Time Constant** | The time required for an instrument to indicate a given percentage of the final reading resulting from an input signal. Synonymous with *lag coefficient*. |
| **Time Delay** | The time required for a specific current or voltage to travel through a circuit. |
| **Time Domain Reflectometry (TDR)** | A method used to characterize and locate faults in metallic cables. In this test, a voltage spike is sent through a conductor and each discontinuity in the conductor path generates a reflected pulse. The reflected pulse and the time difference between initial and reception of the reflected pulse indicate the location of the discontinuity. |
| **Time Series Analysis** | An analytical method to account for the fact that data points taken over time may have an internal structure (such as autocorrelation, trend, or seasonal variation) that should be accounted for. |
| **Time Study** | A work measurement technique consisting of careful time measurement of a task with a time measuring instrument, adjusted for any observed variance from normal effort or pace and to allow adequate time for such items as foreign elements, unavoidable or machine delays, rest to overcome fatigue, and personal needs. Learning or progress effects may also be considered. |
| **Time Study Sheet** | A form for the systematic, detailed recording of element time values, and irregular occurrences, observed during a time study. |

**Time Value of Money** — An economic concept which purports that money available now is more valuable than the same amount of money at some point in the future due simply to its potential earning power, and not inflation as many believe. Used in calculating the present value of money for financial analysis, as well as other purposes.

**Time-Based Maintenance** — Maintenance consisting of periodic (time-based) inspection, service, and cleaning of equipment and replacing of parts to prevent sudden failures and process problems.

**Time-Directed Task** — Periodically scheduled actions aimed directly at failure prevention or retardation.

**Timeline** — A line drawn on a suitable scale (days, months, years, centuries, eons) on which key historical, planned, or projected events and periods are marked in the sequence of their occurrence.

**Timeline Chart** — A graphical representation of a chronological sequence of events.

**Timken Test** — A test to determine the load capability of a lubricant as conducted per ASTM Methods D 2509 (greases) and D 2782 (oils).

**TL 9000** — A quality management standard for the telecommunications industry built on ISO 9000. Its purpose is to define the requirements for the design, development, production, delivery, installation, and maintenance of products and services. Included are cost and performance based measurements that measure reliability and quality performance of the products and services.

**Token** — A unique, structured data object or message that circulates continuously among the nodes of a token ring and describes the current state of the network. Before any node can send a message, it must first wait to control the token.

**Token Passing** — A method of controlling network access through the use of a special signal, called a token, which determines which station is allowed to transmit.

**Token Ring – Network**  A local area network (LAN) formed in a ring (closed loop) topology that uses token passing as a means of regulating traffic on the line.

**Tolerance**  A permissible variation in a characteristic of a product or process, usually shown on a drawing or specification.

**Tolerance Design**  A final stage in product design when allowable component tolerances are tightened if it is expected to avoid more quality loss than the economic loss due to the cost of tolerance tightening.

**Tolerance Limits**  The upper and lower extreme values of the tolerance.

**Toll Manufacturing**  The arrangement through which one company uses its own specialized equipment to make a product for another firm. Synonymous with *contract manufacturing*.

**Tool**  Any instrument or object used by hand to perform work on components, equipment, and facility systems.

**Tool – Method**  A technique, methodology, procedure, or recipe for performing a task or analysis. The term is also applied to software tools (uploading tool).

**Tool Control Program**  A program to manage tools.

**Tool Crib**  A walk-in centralized tool area (crib) assigned to one custodian and shared by multiple users. The crib may be attended, or can be un-attended where control measures are in place to protect the inventory.

**Tool Serviceability**  The physical condition of a tool item that renders it safe and capable for use in its intended purpose. Conditions that render a tool unusable include dullness of cutting surfaces, structural damage due to loose or cracked handles, corrosion, nicks, and cracks.

**Toolbox**  A box, usually compartmentalized, in which tools are kept.

**Tooling and Equipment Supplement**  An interpretation of QS-9000 developed by automobile manufacturers for tooling and equipment suppliers.

**Toolkit**  A collection of methods, techniques, tips and resources which is used as a "how to" reference.

| | |
|---|---|
| **Toolkit – Maintenance** | The set of tools needed by a craftsperson to perform most of the maintenance tasks. |
| **Top Event** | The conceivable, undesired event, in fault tree analysis, to which failure paths of lower level events lead. |
| **Top-Management Commitment** | Participation of the highest level officials in their organization's improvement efforts. Their participation includes:<br>• Establishing and serving on a committee<br>• Establishing policies and goals<br>• Deploying those goals to lower levels of the organization<br>• Providing the resources and training lower levels need to achieve the goals<br>• Participating in quality improvement teams<br>• Reviewing progress organization-wide<br>• Recognizing those who have performed well<br>• Revising the current reward system to reflect the importance of achieving the goals |
| **Topology – Network** | The configuration or layout of a network formed by the connections between devices on a local area network (LAN), or between two or more LANs. |
| **Torque** | Torque is the tendency of a force to cause or change rotational motion of a body. Torque is calculated by multiplying force and distance. The SI unit of torque are Newton-meters. |
| **Torque Wrench** | A wrench that measures the amount of torque being applied to a fastener (nut or bolt). Scales usually read in foot-pounds or Newton-meters. |
| **Torsion** | The twisting of an object due to an applied torque. |
| **Torsional Vibration** | The amplitude modulation of a torque measured in degrees peak-to-peak referenced to the axis of shaft rotation. |
| **Total Acid Number (TAN)** | The quantity of base, expressed in milligrams of potassium hydroxide, which is required to neutralize all acidic constituents present in 1 gram of sample (ASTM Designation D 974). |

**Total Asset Management**  Total asset management is a holistic, inclusive and coordinated approach to asset management. It promotes both a philosophy, and a set of best practices, intended to overcome limiting conditions by coordinating asset-related business processes across multiple business units, integrating asset-related information systems, and adopting best-in-class practices for maintaining and using the information resource.

**Total Available Time**  The theoretical amount of time the production equipment has the potential to produce. Usually the total available time is taken to be:

*24 hours/day x 365 days/year = 8760 hours per year*

**Total Base Number (TBN)**  The quantity of acid, expressed in terms of the equivalent number of milligrams of potassium hydroxide that is required to neutralize all basic constituents present in one gram of a sample (ASTM-Designation D 974).

**Total Cost of Quality**  The aggregate cost of poor quality or product failures including scrap, rework, and warranty costs, as well as expenses incurred to prevent or resolve quality problems (including the cost of inspection).

**Total Downtime**  The amount of time an asset is not capable of running. It is the sum of scheduled and unscheduled downtime.

**Total Dynamic Head (TDH)**  The sum of a pump's dynamic suction head and the dynamic discharge head. Synonymous with *total head*.

**Total Effective Equipment Performance (TEEP)**  A measure of how well an organization is creating value from its assets.

*TEEP = utilization x availability x performance x quality*
*= utilization x OEE*

**Total Employee Involvement**  An empowerment program in which employees are invited to participate in actions and decision making that were traditionally reserved for management.

**Total Float**  The total amount of time that a scheduled activity may be delayed from its start date without delaying the project finish date, or violating a schedule.

| | |
|---|---|
| **Total Productive Maintenance (TPM)** | A maintenance strategy that emphasizes operations and maintenance cooperation. Its goal includes zero defects, zero accidents, zero breakdowns, and an effective workplace design to reduce overall operations and maintenance costs. |
| **Total Quality** | A strategic, integrated system for achieving customer satisfaction that involves all managers and employees and uses quantitative methods to continuously improve an organization's processes. |
| **Total Quality Control (TQC)** | A system that integrates quality development, maintenance, and improvement of the entire organization. It helps a company economically manufacture its product and deliver its services with high quality. |
| **Total Quality Engineering (TQE)** | The discipline of designing quality into the product and manufacturing processes by understanding the needs of the customer and performance capabilities of the equipment. |
| **Total Quality Management (TQM)** | A multifaceted, company-wide approach to improving all aspects of quality and customer satisfaction including fast response and service, as well as product quality. TQM begins with top management and diffuses responsibility to all employees and managers who can have an impact on quality and customer satisfaction. |
| **Total Quality Management Tools** | A set of quality tools, such as benchmarking, quality function deployment, Taguchi methods, statistical process control, design of experiments, and problem-solving methodologies which are used to improve an organization's quality of products and services. |
| **Total System Support** | The composite of all considerations needed to assure the effective and economical support of a system throughout its programmed life-cycle. |
| **Total Work Time Scheduled** | The time in a defined period (1 shift = 8 hours or 480 minutes) during which work is scheduled. |
| **Touch Screen** | A computer screen designed or modified to recognize the location of a touch on its surface. By touching the screen, the user can make a selection or move a cursor. |

| | |
|---|---|
| **Toyota Production System (TPS)** | The production system developed by Toyota Motor Corporation to provide best quality, lowest cost, and shortest lead time through eliminating waste. TPS is based on autonomation and single minute exchange of dies (SMED) principles. |
| **Traceability** | The ability to directly associate a cost with a cost object, as opposed to allocation, which often involves an arbitrary association. The foundation of activity-based cost systems is traceability. |
| **Tracking Filter** | A narrow bandpass filter whose center frequency follows an external synchronizing signal. |
| **Trade** | A specific skill, or set of related skills, in a particular area (millwright, electrician, machinist, boilermaker, carpenter, rigger, etc.). |
| **Trademark** | A logo or insignia that differentiates an organization's goods. |
| **Trade-Off** | Giving up or accepting one advantage, or disadvantage, to gain another that has more value to the decision maker. |
| **Tradesperson** | A skilled worker who normally has completed an apprenticeship program. In some jurisdictions, certain tradespersons must be tested and licensed in their respective trades. |
| **Training** | Instructional activities designed to increase the knowledge, skills, and abilities of persons or teams. |
| **Training Cost** | The costs for formal training that are directed at improving job skills. Training costs should include all employee labor, travel expenses, instructional materials, registration fees, instructor fees, etc. |
| **Training Hours** | The hours for formal training that are directed at improving job skills. |
| **Transaction** | A discrete activity within a computer system, such as an entry of a customer order or an update of an inventory item. Transactions are usually associated with database management, order entry, and other online systems. |

|  |  |
|---|---|
| **Transaction Data** | The finite data pertaining to a given event occurring in a process. For example, the data obtained from testing a machined component (the final product inspection step of the production process). |
| **Transducer** | A device that receives energy or a signal from one system and transmits it to another device, where the input and output energies or signals may be similar or different in form. |
| **Transformation – Organization** | A rapid and dramatic process of total change in values, culture, organization, and procedure to attain significantly higher levels of performance and effectiveness. |
| **Transformer** | A device that converts power from one voltage and current level to another. |
| **Transient Heat Flow** | A thermal condition where the heat flow through a material or system is changing over time. |
| **Transient Vibration** | The short-term vibration of a mechanical system. |
| **Transistor** | An electronic device using a semiconducting material for rectification or amplification of a signal. |
| **Transmissibility (Tr)** | The non-dimensional ratio of response motion/input motion/output in steady-state vibration. |
| **Transverse Sensitivity** | Any output caused by motion that is not along the same axis that a sensor is designed to measure. Synonymous with *cross-axis* and *lateral sensitivity*. |
| **Trap** | A device or piece of equipment for separating one phase from another, as liquid from a gas or water from steam. |
| **Travel Time** | Time required to move material, equipment, personnel, or information from one work or storage area to another. |
| **Tree Diagram** | A management tool that depicts the hierarchy of tasks and subtasks needed to complete an objective. The finished diagram bears a resemblance to a tree. |
| **Trend** | A variable's tendency, over time, to increase, decrease, or remain unchanged. |

| | |
|---|---|
| **Trend Analysis** | A critical examination of trend data, focusing on the long-term directional component. |
| **Trend Control Chart** | A control chart in which the deviation of the subgroup average from an expected trend in the process level is used to evaluate the stability of a process. |
| **Trial** | A single performance of an experiment. |
| **Tribological Wear** | Wear that occurs as a result of relative motion at the surface. |
| **Tribology** | The science and technology of interacting surfaces in relative motion, including the study of lubrication, friction, and wear. |
| **TRIZ** | Developed by Soviet engineer and researcher Genrich Altshuller, it is a systematic approach for creating innovative solutions to technical problems. TRIZ is an acronym for the Russian Теория решения изобретательских задач (Teoriya Resheniya Izobretatelskikh Zadatch) meaning "The theory of inventor's problem solving". |
| **Troubleshooting** | Locating, or isolating, and identifying discrepancies of equipment and determining corrective action. |
| **Truth Table** | A table listing the truth-values of a proposition that result from all the possible combinations of the truth-values of its components. |
| **t-Test** | A test used to assess whether the means of two groups are statistically different from each other. |
| **Turbidity** | The degree of opacity of a fluid caused by the presence of suspended or dissolved particulate or colloidal material. |
| **Turbulent Flow Sampler** | A sampler that contains a flow path in which turbulence is induced in the main stream by abruptly changing the direction of the fluid. |
| **Turnaround** | Planned shutdown of equipment, production line, or process unit to clean, change-out components, make repairs, and if necessary, carry-out capital projects. Synonymous with *outage* and *shutdown*. |

| | |
|---|---|
| **Turnaround Time** | The time required to complete a project or activity. |
| **Turnkey System** | Equipment that is delivered complete, installed, and ready to operate |
| **Turns Ratio – Inventory** | The value of annual maintenance-repair-operations (MRO) stock usage divided by the value of MRO stock inventory. |
| **Turn-to-Turn Insulation** | The insulation between separate wires in each coil on an electric motor. |
| **Twelve Month Rolling Period** | A twelve month period that starts from the current month and goes back twelve months. |
| **Twisted Pair** | A physical communications medium consisting of two copper conductors, each covered with insulation. The two wires are twisted to ensure they are both equally exposed to interference signals in the environment. |
| **Two Bin System** | A type of fixed order system in which inventory is carried in two bins. A replenishment quantity is ordered when the first bin is empty. When the material is received, the reserve bin is refilled and the excess is put into the working bin. |
| **Type** | Generic designation for all equipment of the same kind (conveyors, pumps, haulage trucks, loaders). |
| **Type I Error** | An incorrect decision to reject something (such as a statistical hypothesis or a lot of products) when it is acceptable. |
| **Type II Error** | An incorrect decision to accept something when it is unacceptable. |
| **Types of Redundancies** | The types of redundant configurations that would be used in system design. These configuration types are:<br>• Series<br>• Parallel operating<br>• Standby non-operating |

# U

**U Chart** — A chart of counts per unit.

**Ultraclean** — Having a concentration of fewer than one particle per milliliter that are smaller than 10 microns in size.

**Ultrafiltration** — A fluid purification process in which fluid flows tangentially across a semi-permeable membrane having a highly asymmetric pore structure.

**Ultra-Low Particulate Air Filter** — An air filter medium used in cleanroom applications to filter out particulate sizes as small as 0.125 micron.

**Ultrasonic** — Acoustic frequencies above the range audible to the human ear, or above approximately 20,000 hertz.

**Ultrasonic Testing (UT)** — A non-destructive examination technique for locating defects in a material by passing acoustic energy in the ultrasound range through it. It can be used for pinpointing surface defects, deep defects (e.g. measuring pipe corrosion), detecting pipe leaks, and degradation of bearings. Synonymous with *ultrasonic analysis*.

**Unacceptable Risk** — Exposure to risks that are significant enough to jeopardize an organization's strategy, present dangers to human lives, or represent a significant financial exposure, such that avoidance or mitigation is imperative.

**Uncertainty Allowance** — Allocation of time or money to cover potential occurrence of risk events. Synonymous with *contingency*.

**Underwriters Laboratories (UL)** — An independent testing organization, which examines and tests devices, systems, and materials with particular reference to life, fire, and casualty hazards.

**Undetected Failure** — A potential failure identified in the failure mode and effects analysis (FMEA) where no failure detection method is evident to the operator to make him/her aware of the failure.

| | |
|---|---|
| **Unilateral Tolerance** | A form of tolerance equal to the minimum or maximum performance requirement beyond which a measured performance parameter is considered to have failed. There is an upper bound or a lower bound, but not both. The characteristic must either equal, fall above, or below this limit, as specified, in order to be acceptable. On one side there is conformance, on the other, non-conformance. May be an upper specification limit or lower specification limit. Synonymous with *limit* or *specification limit*. |
| **Uninterruptible Power Supply (UPS)** | A device that provides battery backup when the electrical power fails or drops to an unacceptable voltage level. |
| **Universal Product Code (UPC)** | A barcode symbology that is widely used in the United States and Canada for tracking trade items. |
| **Unloading – Filter** | The release of contaminant that was initially captured by the filter medium. |
| **Unplanned Downtime** | The time equipment is down due to unplanned events such as breakdowns, setups, adjustments, and other documented stoppages. |
| **Unplanned Maintenance** | Maintenance performed without planning which could be related to a breakdown, repair, or corrective work. Unplanned maintenance may be scheduled during the normal schedule cycle. |
| **Unplanned Work** | The amount of unplanned maintenance work that occurred versus the total maintenance hours available. |
| **Unplanned Work Executed** | The unplanned work that has been completed. |
| **Unreliability** | The complement of reliability. |
| **Unscheduled Downtime** | Time an asset is down for repairs or modifications that are not on the weekly maintenance schedule. |
| **Unscheduled Maintenance** | Any maintenance work that has not been included on an approved maintenance schedule prior to its commencement. Synonymous with *unscheduled repairs*. |
| **Unscheduled Work** | Work not on the weekly maintenance schedule. |

| | |
|---|---|
| **Upper Control Limit (UCL)** | Control limit for points above the central line in a control. |
| **Uptime** | The amount of time an asset is actively producing a product or providing a service. It is the actual running time. |
| **Uptime Ratio** | The period of time an asset is functioning and providing a service divided by the total available time. |
| **URL** | An acronym for *u*niversal *r*esource *lo*cator, a naming convention used for finding sites on the World Wide Web. |
| **Useability** | A measure of the ease with which an individual can either operate, or learn how to operate, a system. |
| **Useability Testing** | Empirical testing to measure the usability of a system. |
| **Useful Life** | The maximum length of time that a component can be left in service, before it will start to experience a rapidly increasing probability of failure. |
| **Utilization** | The proportion of available time that an item of equipment is operating. Calculated by dividing equipment operating hours by equipment available hours and generally expressed as a percentage. |
| **Utilization Rate** | A measure of equipment or unit on-stream time expressed as the ratio of actual operating time over planned operating time. Often expressed as a percentage. Synonymous with *utilization factor*. |

# V

**Vacuum Separator**     A separator that utilizes sub-atmospheric pressure to remove certain gases and liquids from another liquid because of their difference in vapor pressure.

**Validation**     The act of confirming a product or service meets the requirements for which it was intended.

**Validity**     The ability of a feedback instrument to measure what it was intended to measure. The degree to which inferences derived from measurements are meaningful.

**Value**     A numerical quantity or relative measure of worth, utility, or importance.

**Value – Financial**     The monetary worth of an item or service, usually determined by the difference between total benefits minus total investment.

**Value Added**     The parts of the process that adds value from the perspective of the external customer.

**Value Adding Process**     Activities that transform input into a customer usable output. The customer can be internal or external to the organization.

**Value Analysis**     The systematic use of techniques to identify the required functions of an item, establish values for those functions, and provide the functions at the lowest overall cost without loss of performance. Synonymous with *value management, value engineering* or *value methodology*.

**Value Chain**     The interdependent matrix of an organization's relationships working together to design, plan, make, ship, sell, and service goods from raw materials to finished products.

**Value Chain Management**     The application of value analysis to every step in a value chain. The goal is to deliver maximum value to the end user for the least possible total cost.

| | |
|---|---|
| **Value Creation** | Creating value in a product or service for the customer. |
| **Value of Stock On Hand** | The current dollar value of the stock in inventory. |
| **Value of Stock Purchased** | The value of the inventory items purchased during a specified time period. |
| **Value Stream** | All activities, both value added and non-value added, required to bring a product from raw material state into the hands of the customer, bring a customer requirement from order to delivery, and bring a design from concept to launch. |
| **Value Stream Mapping** | An analytical tool that helps to visualize and understand the flow of information and material as it makes its way through the process value stream. It identifies non-value-added steps which are waste and need to be removed from the process to improve it. |
| **Values** | The fundamental beliefs that drive organizational behavior and decision making. |
| **Valve** | A device which controls fluid flow direction, pressure, or rate. |
| **Vapor** | A gaseous substance that can be at least partly condensed by moderate cooling or compression. |
| **Vapor Barrier** | A layer of low-permeable material that prevents condensation within building sections. |
| **Vapor Phase** | A substance in the gaseous state, under conditions in which it is capable of being liquefied either by pressure or cooling alone. |
| **Variable** | A physical quantity that is not constant but varies with time. |
| **Variable Cost** | An operation cost that varies directly with a change of one unit in the production volume. |
| **Variable Displacement Pump** | A pump in which the amount of fluid pumped per revolution of the pump's input shaft can be varied while the pump is running. |

| | |
|---|---|
| **Variable Frequency Drive** | A system for controlling the rotational speed of an alternating current (AC) electric motor by controlling the frequency of the electrical power supplied to the motor. A variable frequency drive is a specific type of adjustable-speed drive. Synonymous with *adjustable-frequency drives (AFD), variable-speed drives (VSD), AC drives, microdrives* or *inverter drives.* |
| **Variable Pressure** | The pressure exerted by the vapors released from any material at a given temperature when enclosed in an airtight container. |
| **Variance** | The difference between the expected value and the actual value. |
| **Variance Analysis** | Interpretation of the causes for a difference between actual and some norm, budget, or estimate. |
| **Variance Report** | A report showing the differences between what is expected and what actually occurs, i.e., the variance. |
| **Variation** | A change in data, characteristic, or function caused by one of four factors:<br>1. Special causes<br>2. Common causes<br>3. Tampering<br>4. Structural variation |
| **Varnish** | A thin, insoluble, non-wipeable film deposit, in lubrication, occurring on interior parts, resulting from the oxidation and polymerization of fuels and lubricants. Can cause sticking and malfunction of close-clearance moving parts. Similar to, but softer, than lacquer. |
| **Vector** | A quantity which has both magnitude and direction (or phase). |
| **Velocity** | Rate of change of displacement with time, usually along a specified axis. It may refer to angular motion as well as uniaxial motion. |
| **Vendor** | A supplier or distributor of commonly available goods or services when requirements and specifications are well defined. |

| | |
|---|---|
| **Vendor Managed Inventory** | A supply strategy in which raw materials, components, spare parts, and other inventories are stocked by the manufacturer or supplier specifically designated for a customer. Supplying the items on-demand or through routine deliveries when customers' stocks are low. Synonymous with *vendor stocking program*. |
| **Venn Diagram** | A schematic representation of the universal set and its subsets. A rectangle is usually used to represent the universal set, points to designate elements of the set, and circles to depict subsets of the universal set. Synonymous with *Euler Diagram*. |
| **Venturi Tube** | A tube, inserted in a line, whose internal surface consists of two truncated cones connected at the small ends by a short cylinder (the throat). As the velocity of the flow of the fluid increases in the throat, the pressure decreases. The tube is used to measure the quantity of fluid flowing, or by joining a branch tube at the throat to produce suction. |
| **Verbal Orders** | A means of assigning emergency work when reaction time does not permit preparation of a work order document. |
| **Verification** | The act of determining whether products and services conform to specific requirements. |
| **Verification Readiness Review** | A review of design and build status to determine whether the system is ready to be operated for system functional and performance checks. |
| **Vertical Carousels** | Vertical carousels consist of shelving layers of pans or trays mounted on a vertical revolving system. |
| **Vertical Deployment** | A term that denotes all levels of management are involved in the organization's quality efforts. |
| **Vibration** | Mechanical oscillation or motion about a reference point. |
| **Vibration Analysis** | The analysis of vibration monitoring data to monitor characteristic changes in rotating machinery caused by imbalance, misalignment, bent shaft, mechanical looseness, faults in gear drives, defects in rolling-element bearings, and/or defects in sleeve bearings. |

| | |
|---|---|
| **Vibration Machine** | A device which produces controlled and reproducible mechanical vibration for the vibration testing of mechanical systems, components, and structures. Synonymous with *exciter* or *shaker*. |
| **Vibration Meter** | An apparatus (usually an electronic amplifier, detector, and readout meter) for measuring electrical signals from vibration sensors. May display displacement, velocity and/or acceleration. |
| **Vibration Monitoring** | A technology used to determine equipment condition and potentially predict failure. Equipment can be monitored using instrumentation such as vibration analysis equipment or the human senses. |
| **Vicarious Learning** | A component of social learning theory involving our ability to learn new behaviors and/or assess their probable consequences by observing others. |
| **Vicious Cycle** | A situation where the solution to a circumstance creates another problem in a chain that makes the original problem worse. An example is when a plant attempts to save labor by reducing preventive maintenance. The incidence of equipment failure rises requiring more labor and leaving no time for performing preventive maintenance. The plant finally saves no labor and has worse equipment reliability. |
| **Video Graphics Array (VGA)** | A video adapter that duplicates all the video modes of the EGA (Enhanced Graphics Adapter) and adds several more. |
| **Virtual Team** | Remotely situated individuals affiliated with a common organization, purpose, or project who conduct their joint effort via electronic communication. |
| **Viscometer** | An apparatus for determining the viscosity of a fluid. Also spelled viscosimeter. |
| **Viscosity** | A measurement of a fluid's resistance to flow. The common metric unit of absolute viscosity is the poise, which is defined as the force in dynes required to move a surface one square centimeter in area past a parallel surface at a speed of one centimeter per second, with the surfaces separated by a fluid film one centimeter thick. |

| | |
|---|---|
| **Viscosity Grade** | Any of a number of systems which characterize lubricants according to viscosity for particular applications, such as industrial oils, gear oils, automotive engine oils, automotive gear oils, and aircraft piston engine oils. |
| **Viscosity Index (VI)** | A commonly used measure of a fluid's change of viscosity with temperature. The higher the viscosity index, the smaller the relative change in viscosity with temperature. |
| **Viscosity Index Improvers** | Additives that increase the viscosity of the fluid throughout its useful temperature range. Such additives are polymers that possess thickening power as a result of their high molecular weight and are necessary for formulation of multi-grade engine oils. |
| **Viscosity Modifier** | A lubricant additive, usually a high molecular weight polymer, which reduces the tendency of an oil's viscosity to change with temperature. |
| **Viscous** | Possessing viscosity. Frequently used to imply high viscosity. |
| **Visibility Systems** | Visual systems on the plant floor, design areas, and elsewhere that enable anyone familiar with the work to understand its status and condition at a glance, or to respond to work priorities. This can be done with standard layouts, signal lights, kanban systems, or other methods. The distinguishing feature is that communication is rapidly executed by line of sight. |
| **Vision** | An overarching statement of the way an organization wants to be. An ideal state of being at a future point. |
| **Visual Control** | The use of easy-to-read indicators to show equipment status and performance (red, yellow, or green gauge markings; normal reading zone indicators; color-coded oil cans and filler caps). |
| **Visual Workplace** | The use of visual displays to relay information to employees and guide their actions. The workplace is usually set up with signs, labels, color-coded marking etc., so that anyone unfamiliar with the assets or the process can readily identify what is going on, understand the process, know what's being done correctly and what is out of place. |

**Vital Few, Useful Many** — A term used by Joseph M. Juran to describe his use of the Pareto principle, which he first defined in 1950 (the principal was used much earlier in economics and inventory control methodologies). The principle suggests most effects come from relatively few causes. That is, 80% of the effects come from 20% of the possible causes. The 20% of the possible causes are referred to as the *vital few*, the remaining causes are referred to as the *useful many*. When Juran first defined this principle, he referred to the remaining causes as the *trivial many*, but realizing that no problems are trivial in quality assurance, he changed it to *useful many*.

**Voice Of the Customer (VOC)** — The expressed requirements and expectations of customers relative to products or services, as documented and disseminated to the members of the providing organization.

**Voice Recognition** — Computerized systems capable of recognizing or synthesizing human voices. Such systems capture verbalized data for quality-control or inventory-tracking purposes (often when operators' hands are busy), recognize spoken commands that activate equipment, and convert computer data into audible information.

**Volatility** — The degree and rate at which a liquid will vaporize under given conditions of temperature and pressure. When liquid stability changes, this property is often reduced in value.

**Voltage** — The force that causes a current to flow in an electrical circuit.

**Volute** — The casing surrounding the impeller in a centrifugal pump that collects the liquid discharged from the impeller.

# W

**Wage**
Payment for work performed. Usually applied to blue collar work as opposed to salary applied to white-collar work.

**Waiting Time**
The time when an operator, maintenance person, or a machine waits for service, parts, inspection, instructions, and other causes. Synonymous with *delay time*.

**Walk-Through**
Rehearsal of an operational procedure by simulating the execution of all its steps, except those that are high risk or prohibitively expensive.

**Warehouse**
A storage facility where materials and spare parts are stocked.

**Warehouse Management Systems**
Software that integrates mechanical and human performed activities with an information system to effectively manage warehouse business processes and direct warehouse activities. Systems that automate receiving, put away, picking, and shipping in warehouses, and that can prompt workers to do inventory cycle counts. Most support radio-frequency communications, allowing real-time data transfer between the system and warehouse personnel.

**Warranty**
Legally binding assurance (which may or may not be in writing) that a product or service is, among other things:

- Fit for use as represented
- Free from defective material and workmanship
- Meets statutory and/or other specifications

A warranty describes the conditions and period during which the producer or vendor will repair, replace, or otherwise compensate for the defective item without cost to the buyer or user. Often it also delineates the rights and obligations of both parties in case of a claim or dispute.

| | |
|---|---|
| **Waste** | Any activity that consumes resources and produces no added value to the product or service a customer receives. There are seven types of manufacturing waste:<br>1. Production over immediate demand<br>2. Excess work in progress and finished goods inventories<br>3. Scrap, repairs and rejects<br>4. Unnecessary motion<br>5. Excessive processing<br>6. Wait time<br>7. Unnecessary transportation |
| **Water Hammer** | Pressure surges that result from a sudden velocity reduction of a flowing liquid in a contained system. Synonymous with *hydraulic hammer*. |
| **Wave** | Nature's mechanism for transporting energy without transporting matter. |
| **Waveform** | A presentation or display of the instantaneous amplitude of a signal as a function of time, as on an oscilloscope or oscillograph. |
| **Waveform Analysis** | The determination of the amplitude and phase of the components of a complex waveform, either mathematically or by means of electronic instruments. |
| **Wear** | Attrition or rubbing away of the surface of a material as a result of mechanical action. |
| **Wear Particle Analysis** | Wear particle analysis (test) emphasizes the detection and analysis of machine anomalies, which are the symptoms of failure. The oil serves as the messenger of the information on the health of the machine. The presence of abnormal number of wear particles, their size, shape, color, orientation, etc., define the cause, source, and severity of the condition. |
| **Wearout** | The process that results in an increase of the failure rate or probability of failure with increasing age. |
| **Wearout Stage** | The final stage before failure of a component that is wearing out. It is characterized by a sharp rise in the failure rate. |

| | |
|---|---|
| **Weekly Schedule** | The list of maintenance work to be done in the week. It is usually finalized three to four days before the start of the work week. |
| **Weep Hole** | A small hole in an orifice plate to prevent liquid from accumulating upstream. |
| **WeiBayes Distribution** | A special case of the Weibull failure distribution where the Beta (slope) value can be defined by the user. |
| **Weibull Distribution** | A failure distribution that is used in reliability analysis to model life distributions. By adjusting the beta factor, or shape parameter, of the Weibull distribution, it can be made to model a decreasing, constant, or increasing hazard rate. |
| **Weibull Plot** | A graphical technique for determining if a data set comes from a population that would logically be fit by a 2-parameter Weibull distribution. The Weibull plot has special scales that are designed so that if the data do in fact follow a Weibull distribution, the points will be linear (or nearly linear). |
| **Weighed Voting** | A way to prioritize a list of issues, ideas, or attributes by assigning points to each item based on its relative importance. |
| **Weight** | That property of an object that can be weighed, as on a scale. The gravitational force on an object. |
| **What-If Analysis** | The process of evaluating alternate strategies by answering the consequences of changes to forecasts, manufacturing plans, inventory levels, etc. |
| **Whiskers** | Electrically conductive, crystalline structures of tin that sometimes grow from surfaces where tin, especially electroplated tin, is used as a final finish. Whiskers that bridge closely spaced circuit elements can cause short circuits which can lead to electronic system failures. Synonymous with *tin whiskers*. |
| **White Noise** | Noise that contains all sound frequencies. It is often used for sound masking in offices and work areas. |

| | |
|---|---|
| **White Random Vibration** | The broad-band random vibration in which the power spectral density (PSD) - or auto spectral density (ASD) - is constant over a broad frequency range. |
| **White-Collar Worker** | Usually associated with the description of an employee on the office, clerical, sales, technical and/or professional staffs. |
| **Wicking** | The vertical absorption of a liquid into a porous material by capillary forces. |
| **Wide Area Network (WAN)** | The interconnection of computers and LANs over a long distance to create a network for the exchange and sharing of data and resources. Long distance may be several miles or across state or country borders. The Internet is the largest WAN. |
| **Wien's Displacement Law** | The law that describes the relationship between the temperature of a blackbody and the peak wavelength of radiation it gives off. |
| **Wilcoxon Mann-Whitney Test** | Used to test the null hypothesis that two populations have identical distribution functions against the alternative hypothesis that the two distribution functions differ only with respect to location (median), if at all. It does not require the assumption that the differences between the two samples are normally distributed. In many applications, it is used in place of the two sample t-test when the normality assumption is questionable. This test can also be applied when the observations in a sample of data are ranks, that is, ordinal data rather than direct measurements. |
| **Winding** | The wrapping coils of copper wire around a core, usually of steel, which are fond in motors. In an AC induction motor, the primary winding is a stator consisting of wire coils inserted into slots within steel laminations. The secondary winding of an AC induction motor is usually not a winding, but rather a cast rotor assembly. In a permanent magnet DC motor, the winding is the rotating armature. |
| **Win-Lose** | Outcome of conflict resolution that typically makes use of the power available to each party and treats conflict as a zero-sum game. |

| | |
|---|---|
| **Win-Win** | Outcome of conflict resolution that results in both parties being better off. Focuses on the objectives of both parties and the ways to meet those objectives while resolving the issue at hand. |
| **WIP Turn Rate** | A measure of the speed with which work-in-process (WIP) moves through a plant. Typically calculated by dividing the value of total annual shipments at plant cost by the average WIP value. |
| **Wiring Diagram** | The graphic representation of all circuits and device elements of an electrical system. The representation may be physical or schematic. |
| **Work Acceptance** | Work is considered accepted when it is conducted, documented, and verified according to acceptance criteria. |
| **Work Authorization** | Permission for specific work to be performed during a specific time period. |
| **Work Breakdown Structure (WBS)** | A hierarchically-structured grouping of project elements that organizes and defines the total scope of a project. Each descending level is an increasingly detailed definition of project component. |
| **Work Control** | The function of controlling asset task executions by planning, scheduling, and coordinating activities so that work performers are provided with job plans, schedules, materials, tools, and supporting services to accomplish tasks effectively. |
| **Work Controller** | The role of approving, prioritizing, coding, establishing status, routing, and coordinating work requests and work orders so that planning and scheduling can be effectively performed prior to task execution. |
| **Work Flow Analysis (WFA)** | A method of analyzing the progress or "flow" of work within a system or process. It categorizes process tasks into value-added, value-assisting, and non-value added. |
| **Work Force** | All the employees of an organization. |
| **Work Groups** | The organization of workers functionally and/or administratively as a unit with a unique and definable mission. |

| | |
|---|---|
| **Work in Process (WIP)** | The amount or value of all materials, components, and subassemblies representing partially completed production and anything between the raw material/purchased component stage and finished-goods stage. Value should be calculated at plant cost, including material, direct labor, and overhead. Synonymous with *work in process inventory*. |
| **Work Load** | The essential work to be performed by maintenance and the conversion of this data into a work force of the proper size and craft composition to ensure that the program is carried out effectively. |
| **Work Order (WO)** | The prime document used by the maintenance function to manage maintenance tasks. It may include such information as a description of the work required, the task priority, the job procedure to be followed, the parts, materials, tools and equipment required to complete the job, the labor hours, costs and materials consumed in completing the task, as well as key information on failure causes, what work was performed, etc. |
| **Work Order Parts Kitting** | The purchase, withdrawal, packaging, and staging of parts for each individual work order. |
| **Work Order System** | A communications system by which maintenance is requested, classified, planned, scheduled, assigned, and controlled. |
| **Work Package** | Deliverable at the lowest level of a work breakdown structure. May be divided into activities and used to identify and control work flow. |
| **Work Plan** | An information packet provided to the work performer which contains job specific requirements such as task descriptions sequenced in steps, job specific and safety permits/procedures, drawings, materials, and tools required to perform the job effectively. |
| **Work Request (WR)** | A simple request for maintenance service or work requiring no planning or scheduling but usually a statement of the problem. Usually precedes the issuance of a work order. |
| **Work Results** | Outcome of activities performed to accomplish a task or project. |

| | |
|---|---|
| **Work Rules** | Workplace requirements established to maintain discipline, prevent injuries and accidents, and maintain productivity. |
| **Work Sampling** | A procedure for observing work activities at random intervals of time from which statistical inference may be made relative to the entire scope of work involved. |
| **Work Team** | A team comprising members from one work unit. Synonymous with *natural team*. |
| **Workable Backlog** | The total man-hours estimated on repair work orders that are ready to work (all parts and materials available) divided by the average man-hours available per week (typically expressed in crew-weeks). |
| **Working Capital** | Current assets less current liabilities. A measure of liquidity. |
| **Working Foreman** | The role of supporting a supervisor by leading work performers in task executions, clarifying task assignments and duties, providing technical and administrative guidance, and coordinating and de-conflicting interactions between supporting crews. |
| **Working Time – Employee** | The number of hours an employee is scheduled to be on the job during a day or week (8 hours per day or 40 hours per week). |
| **Working Time – Equipment** | The number of hours that equipment or the plant is expected to operate. Calculated by subtracting lost time from production adjustments, or periodic servicing from the calendar time. |
| **Workload** | The number of labor hours needed to carry out a maintenance program, including all scheduled and unscheduled work and maintenance support of project work. |
| **Workspace** | The physical area in which an individual performs some duty or task. |
| **Workstation** | An area used in a manufacturing process to perform a series of functional tasks, usually associated with a single operator. Also refers to an individual work space (cubicle). |

| | |
|---|---|
| **Workstation – Computer** | A stand-alone computer used in computer-aided design and other calculational or graphically demanding applications. |
| **World Wide Web (www)** | The total set of interlinked hypertext documents residing on HTTP servers all around the world. |
| **World-Class** | Leading performance in a process independent of industry, function, or location. |
| **World-Class Manufacturer** | A somewhat arbitrary designation that can be supported by performance results related to various manufacturing metrics (world-class metrics may vary from one industry to another). Typically, it denotes "best in class" producers on a worldwide basis. In the broadest sense, world-class manufacturers are those perceived to deliver the greatest value at a given price level. |
| **Wrench Time** | A measure of the time a maintenance craft worker spends applying physical effort, or troubleshooting, in the accomplishment of assigned work. |
| **WYSIWYG** | A screen format that displays information (text, graphics, etc.) very similar to the final output. An acronym for *What You See Is What You Get*. |

# X

**X Bar Chart** — Chart of the sample averages.

**X Metric** — Measurable independent input factors that determine the outcome (Y) of a process. Six Sigma focuses on process X's to improve process output.

**X-Axis** — In a plane Cartesian coordinate system, the axis along which the abscissa is measured and from which the ordinate is determined. In a three-dimensional Cartesian coordinate system, the axis along which the values of x are measured and at which y and z are equal to zero.

**XOR Gate** — A logic gate in which an output occurs if exactly one input event occurs.

# Y

**Y = f(x)** — Complex transfer function that describes how process elements are related. Process outputs (Ys) are a function of process inputs (Xs).

**Yaw** — Rotation about the vertical axis.

**Y-Axis** — In a plane Cartesian coordinate system, the axis along which the ordinate is measured and from which the abscissa is determined. In a three-dimensional Cartesian coordinate system, the axis along which the values of y are measured and at which x and z are equal to zero.

**Yield** — The percentage of defect-free units produced.

**Young's Modulus (E)** — The modulus of elasticity for pure tension with no other stress activity. In most metals, it is practically the same in compression. Denoted by E it is the tensile (or compressive) stress per unit of linear strain or tensile stress intensity/tensile strain.

# Z

**Z-Axis** — In a three-dimensional Cartesian coordinate system, the axis, perpendicular to the x and y axes, along which the values of z are measured, and at which x and y are equal to zero.

**ZDDP** — Zinc dialkyldithiophosphate, an anti-wear additive found in many types of hydraulic and lubricating fluids.

**Zero Defects** — A performance standard and methodology that states if people commit to watching details and avoiding errors, they can move closer to the goal of achieving defect-free products and services.

**Zero-Base Budgeting** — A budget approach in which responsibility centers start with zero in preparing their budget requests and must justify the contributions of each of their activities to organizational goals.

**Zero-G Drift** — The amount that a sensor's signal shifts over some temperature range.

**Zero-G Output** — The output that is read when the sensor is not accelerating.

**Zero-to-Peak** — Half of the peak-to-peak value.

# Acronyms & Initialisms

| | |
|---|---|
| 2D | Two-Dimensional |
| 3D | Three-Dimensional |
| AACE | Association for the Advancement of Cost Engineering |
| ABC | Activity-Based Costing |
| AC | Alternating Current |
| ADA | Americans with Disabilities Act |
| ADAAA | Americans with Disabilities Act Amendments Act of 2008 |
| AFE | Association for Facilities Engineering |
| $A_i$ | Availability (Inherent) |
| AI | Artificial Intelligence |
| AIAG | Automotive Industry Action Group |
| AISI | American Iron and Steel Institute |
| ALARP | As Low As Reasonably Practical |
| ALT | Accelerated Life Testing |
| ANOM | ANalysis Of Means |

| | |
|---|---|
| **ANOVA** | ANalysis Of VAriance |
| **ANSI** | American National Standards Institute |
| **$A_o$** | Availability (Operational) |
| **API** | American Petroleum Institute |
| **APQC** | American Productivity and Quality Council |
| **AQL** | Acceptable Quality Level |
| **ASCII** | American Standard Code for Information Interchange |
| **ASD** | Auto Spectral Density |
| **ASME** | American Society for Mechanical Engineers |
| **ASNT** | American Society for Nondestructive Testing |
| **ASQ** | American Society for Quality |
| **ASSE** | American Society of Safety Engineers |
| **ASTM** | American Society of Testing and Materials |
| **AU** | Asset Utilization |
| **BHAG** | Big Hairy Audacious Goal |
| **BIT** | Built-In Test |
| **BITE** | Built-In Test Equipment |
| **BLS** | Bureau of Labor Statistics |

| | |
|---|---|
| **BOK** | Body Of Knowledge |
| **BOM** | Bill Of Materials |
| **BPM** | Business Process Management |
| **BPM** | Business Process Modeling |
| **BPMN** | Business Process Modeling Notation |
| **BPR** | Business Process Reengineering |
| **BSI** | British Standards Institute |
| **CAD** | Computer Aided Design |
| **CAE** | Computer Aided Engineering |
| **CAI** | Computer Aided Instruction |
| **CAM** | Computer Aided Manufacturing |
| **CAP** | Certified Automation Professional |
| **CAPP** | Computer Aided Process Planning |
| **CASE** | Computer Aided Software Engineering |
| **CAV** | Current Asset Value |
| **CBM** | Condition Based Maintenance |
| **CBT** | Certified Balancing Technician |
| **CBT** | Computer-Based Training |

| | |
|---|---|
| **CCB** | Configuration Control Board |
| **CCST** | Certified Control Systems Technician |
| **CE** | Concurrent Engineering |
| **CEN** | European Committee for Standardization |
| **CFD** | Computational Fluid Dynamics |
| **CFR** | Code of Federal Regulations (US) |
| **CFT** | Cross Functional Team |
| **cGMPs** | Current Good Manufacturing PracticeS |
| **CI** | Continuous Improvement |
| **CIM** | Computer Integrated Manufacturing |
| **CIMM** | Certified Industrial Maintenance Mechanic |
| **CL** | CenterLine |
| **CL** | Confidence Level |
| **CLCA** | Closed-Loop Corrective Action |
| **CM** | Condition Monitoring |
| **CM** | Corrective Maintenance |
| **CMII** | Configuration Management |
| $C_{mk}$ | Machine Capability Index |

| | |
|---|---|
| **CMMS** | Computerized Maintenance Management System |
| **CMRP** | Certified Maintenance and Reliability Professional |
| **CNC** | Computer Numerical Control |
| **COGM** | Cost Of Goods Manufactured |
| **COGS** | Cost Of Goods Sold |
| **CoP** | Community of Practice |
| **COPQ** | Cost Of Poor Quality |
| **COQ** | Cost Of Quality |
| **COTS** | Commercial Off The Shelf |
| $C_p$ | Capability Index |
| **CPE** | Certified Plant Engineer |
| **CPI** | Chemical Process Industries |
| **CPI** | Consumer Price Index |
| **CPI** | Continuous Process Improvement |
| **CPI** | Cost Performance Index |
| $C_{pk}$ | Capability Performance Index |
| **CPM** | Critical path method |
| **CPMM** | Certified Plant Maintenance Manager |

| | |
|---|---|
| **CPS** | Certified Plant Supervisor |
| **CPS** | Cycles Per Second |
| **CPU** | Central Processing Unit |
| **CQA** | Certified Quality Auditor |
| **CQE** | Certified Quality Engineer |
| **CQI** | Certified Quality Inspector |
| **CQI** | Continuous Quality Improvement |
| **CQM** | Certified Quality Manager |
| **CQT** | Certified Quality Technician |
| **CRE** | Certified Reliability Engineer |
| **CRM** | Customer Relationship Management |
| **CRT** | Cathode Ray Tube |
| **CRV** | Current Replacement Value |
| **CSA** | Canadian Standards Association |
| **CSQE** | Certified Software Quality Engineer |
| **CSSBB** | Certified Six Sigma Black Belt |
| **CSSGB** | Certified Six Sigma Green Belt |
| **CTQ** | Critical-To-Quality |

| | |
|---|---|
| **CVA** | Certified Vibration Analyst |
| **CVC** | Certified Vibration Consultant |
| **DC** | Direct Current |
| **DCF** | Discounted Cash Flow |
| **DCS** | Distributed Control System |
| **DFA** | Design For Assembly |
| **DFEMA** | Design Failure Mode and Effects Analysis |
| **DFM** | Design For Manufacturability |
| **DFMt** | Design For Maintainability |
| **DFR** | Design For Reliability |
| **DFSS** | Design For Six Sigma |
| **DMADV** | Define, Measure, Analyze, Design and Verify |
| **DMAIC** | Define, Measure, Analyze, Improve, Control |
| **DoD** | Department of Defense |
| **DoE** | Design of Experiments |
| **DPMO** | Defects Per Million Opportunities |
| **DSP** | Digital Signal Processing |
| **DSS** | Decision Support System |

| | |
|---|---|
| **DTLCC** | Design To Life Cycle Cost |
| **EAC** | Estimate At Completion |
| **EAM** | Enterprise Asset Management system |
| **EC** | Engineering Change |
| **EDI** | Electronic Data Interchange |
| **EDP** | Electronic Data Processing |
| **EFNMS** | European Federation of National Maintenance Societies |
| **EFQM** | European Foundation for Quality Management |
| **EI** | Enterprise Integration |
| **EIA** | Electronic Industry Association |
| **EMF** | ElectroMotive Force |
| **EN** | European Norm |
| **EOQ** | Economic Order Quantity |
| **EOY** | End Of Year |
| **EPA** | Environmental Protection Agency |
| **EPRI** | Electric Power Research Institute |
| **ERP** | Enterprise Resource Planning |
| **ERV** | Equipment Replacement Value |

| | |
|---|---|
| **ERV** | Estimated Replacement Value |
| **ETA** | Event Tree Analysis |
| **ETC** | Estimate To Complete |
| **EVA** | Economic Value Added |
| **FAQ** | Frequently Asked Questions |
| **FDA** | Food and Drug Administration |
| **FEA** | Finite Element Analysis |
| **FEED** | Front-End Engineering Design |
| **FEL** | Front-End Loading |
| **FFT** | Fast Fourier Transform |
| **FIFO** | First-In-First-Out |
| **FIN** | Functional Identification Number |
| **FMA** | Failure Mode Analysis |
| **FMEA** | Failure Mode and Effects Analysis |
| **FMECA** | Failure Mode, Effects, and Criticality Analysis |
| **FMS** | Flexible Manufacturing System |
| **FOB** | Free On Board |
| **FPA** | Focal Plane Array |

| | |
|---|---|
| **FRACAS** | Failure Reporting, Analysis, and Corrective Action System |
| **FSA** | Fluid Sealing Association |
| **FTA** | Fault Tree Analysis |
| **FTIR** | Fourier Transform Infrared Spectroscopy |
| **G&A** | General And Administrative |
| **GAAP** | Generally Accepted Accounting Principles |
| **GD&T** | Geometric Dimensioning And Tolerancing |
| **GERT** | Graphical Evaluation and Review Technique |
| **GFCI** | Ground Fault Circuit Interrupter |
| **GIDEP** | Government Industry Data Exchange Program |
| **GIGO** | Garbage In, Garbage Out |
| **GR&R** | Gauge Repeatability And Reproducibility |
| **GTAW** | Gas Tungsten Arc Welding |
| **GUI** | Graphical User Interface |
| **HA** | Hazard Analysis |
| **HACCP** | Hazard Analysis and Critical Control Point |
| **HALT** | Highly Accelerated Life Test |
| **HAZOP** | HAZard and OPerability Study |

| | |
|---|---|
| **HEPA** | High Efficiency Particulate Air (filters) |
| **HiPot** | HIgh POTential |
| **HMI** | Human Machine Interface |
| **HP** | Horse Power |
| **HR** | Human Resources |
| **HTML** | HyperText Markup Language |
| **HVAC** | Heating, Ventilating, and Air Conditioning |
| **HVI** | High Viscosity Index |
| **I&C** | Instrument And Controls |
| **I/O** | Input/Output |
| **IC** | Integrated Circuit |
| **ICML** | International Council for Machinery Lubrication |
| **ID** | Inside Diameter |
| **IEC** | International Electrotechnical Commission |
| **IEEE** | Institute of Electrical and Electronic Engineers |
| **IEMD** | Institute of Electric Motor Diagnostics |
| **IFOV** | Instantaneous Field Of View |
| **IIE** | Institute of Industrial Engineers |

| | |
|---|---|
| **INCOSE** | International Council Of Systems Engineering |
| **IR** | InfraRed |
| **IRP** | Insulation Resistance Profile |
| **IRR** | Internal Rate of Return |
| **IS** | Information Systems |
| **ISA** | International Society of Automation |
| **ISI** | In-Service Inspection |
| **ISO** | International Organization for Standardization |
| **IT** | Information Technology |
| **JIPM** | Japanese Institute of Plant Maintenance |
| **JIT** | Just-In-Time |
| **JPEG** | Joint Photographic Experts Group |
| **KISS** | Keep It Simple and Specific |
| **KPI** | Key Performance Indicator |
| **LAN** | Local Area Network |
| **LCC** | Life Cycle Cost |
| **LCD** | Liquid Crystal Display |
| **LCL** | Lower Control Limit |

| | |
|---|---|
| **LED** | Light-Emitting Diode |
| **LEED** | Leadership in Energy and Environmental Design |
| **LEED AP O+M** | LEED Accredited Professional in Operation and Maintenance |
| **LEL** | Lower Explosive Limit |
| **LIFO** | Last-In-First-Out |
| **LLA** | Laboratory Lubrication Analyst |
| **LNG** | Liquefied Natural Gas |
| **LOE** | Lost Opportunity Event |
| **LORA** | Level Of Repair Analysis |
| **LRU** | Line Replacement Unit |
| **LTA** | Logic Tree Analysis |
| **LVI** | Low Viscosity Index |
| **M&R** | Maintenance And Reliability |
| **MAPI** | Machinery and Allied Products Institute |
| **MBB** | Master Black Belt |
| **MBE** | Management By Example |
| **MBNQA** | Malcolm Baldrige National Quality Award |
| **MBO** | Management By Objectives |

| | |
|---|---|
| **MBWA** | Management By Walking Around |
| **MCA** | Motor Current Analysis |
| **MCSA** | Motor Current Signature Analysis |
| **MDT** | Mean DownTime |
| **MEMS** | MicroElectroMechanical Structures |
| **MES** | Manufacturing Execution System |
| **MFA** | Motor Flux Analysis |
| **MI** | Mechanical Integrity |
| **MIL** | MILitary |
| **MIL-HDBK** | U.S. MILitary HanDBooK |
| **MIL-STD** | U.S. MILitary STandarD |
| **MIMOSA** | Machinery Information Management Open Systems Alliance |
| **MIS** | Management Information System |
| **MKSA** | Meter, Kilogram, Second, Ampere |
| **MLA** | Machine Lubricant Analyst |
| **MLT** | Machine Lubrication Technician |
| **MOC** | Management Of Change |
| **MRO** | Maintenance, Repair, and Operations |

| | |
|---|---|
| **MRP** | Materials Requirements Planning |
| **MRP II** | Manufacturing Resource Planning |
| **MSDS** | Material Safety Data Sheet |
| **MTBE** | Mean Time Between Events |
| **MTBF** | Mean Time Between Failures |
| **MTBM** | Mean Time Between Maintenance |
| **MTBR** | Mean Time Between Repair |
| **MTTF** | Mean Time To Failure |
| **MTTR** | Mean Time To Repair |
| **MUX** | MUltipleXer |
| **NACFAM** | National Council For Advanced Manufacturing |
| **NAICS** | North American Industry Classification System |
| **NAME** | North American Maintenance Excellence (award) |
| **NAS** | National Aerospace Standard |
| **NASA** | National Aeronautics and Space Administration |
| **NC** | Numerical Control |
| **NDE** | Non-Destructive Examination |
| **NDT** | NonDestructive Testing |

| | |
|---|---|
| **NEC** | National Electrical Code |
| **NEMA** | National Electrical Manufacturers Association |
| **NFPA** | National Fire Protection Association |
| **NIH** | Not Invented Here Syndrome |
| **NIST** | National Institute of Standards and Technology |
| **NPRA** | National Petroleum Refiners Association |
| **NPSH** | Net Positive Suction Head |
| **NPV** | Net Present Value |
| **NRC** | Nuclear Regulatory Commission |
| **NSPE** | National Society of Professional Engineers |
| **O&M** | Operations And Maintenance |
| **OBM** | Operator Based Maintenance |
| **OCR** | Optical Character Recognition |
| **OD** | Outside Diameter |
| **ODR** | Operator Driven Reliability |
| **OEE** | Overall Equipment Effectiveness |
| **OEM** | Original Equipment Manufacturer |
| **OH** | OverHead |

| | |
|---|---|
| **OJT** | On the Job Training |
| **OPC** | OPen Connectivity |
| **OSHA** | Occupational Safety and Health Administration |
| **P&ID** | Piping And Instrumentation Diagram |
| **P.E.** | Professional Engineer (U.S.) |
| **P. Eng.** | Professional ENGineer (Canada) |
| **PBS** | Project Breakdown Structure |
| **PC** | Personal Computer |
| **PCA** | Physical Configuration Audit |
| **PCB** | PolyChlorinated Biphenyls |
| **PCB** | Printed Circuit Board |
| **PDA** | Personal Digital Assistant |
| **PDCA** | Plan-Do-Check-Act |
| **PDF** | Portable Document Format |
| **PDM** | Precedence Diagramming Method |
| **PdM** | Predictive Maintenance |
| **PDPC** | Process Decision Program Charts |
| **PE** | PiezoElectric |

| | |
|---|---|
| **PERT** | Project Evaluation and Review Technique |
| **pf** | Power Factor |
| **PFD** | Process Flow Diagram |
| **PHA** | Process Hazard Analysis |
| **PID** | Proportional plus Integral plus Derivative (Control) |
| **PLC** | Programmable Logic Controller |
| **PM** | Preventive Maintenance |
| **PMBOK** | Project Management Body Of Knowledge |
| **PMI** | Project Management Institute |
| **PMP** | Project Management Professional |
| **PO** | Purchase Order |
| **PPE** | Personal Protective Equipment |
| **PPM** | Parts Per Million |
| **PPP** | Pre-Project Planning |
| **PR** | PiezoResistive |
| **PRA** | Probabilistic Risk Assessment |
| **PRV** | Plant Replacement Value |
| **PSA** | Probabilistic Safety Assessment |

| | |
|---|---|
| **PSD** | Power Spectral Density |
| **PSM** | Process Safety Management |
| **QA** | Quality Assurance |
| **QBS** | Qualifications-Based Selection |
| **QC** | Quality Control |
| **QEDS** | Quality, Environment, Dependability, and Statistics |
| **QFD** | Quality Function Deployment |
| **QM** | Quality Management |
| **QMS** | Quality Management System |
| **QRB** | Quality Review Board |
| **R&D** | Research And Development |
| **R&M** | Reliability And Maintenance |
| **RAM** | Random Access Memory |
| **RAM** | Reliability, Availability, and Maintainability |
| **RAMS** | Reliability, Availability, Maintainability, and Safety |
| **RAV** | Replacement Asset Value |
| **RBD** | Reliability-Based Design |
| **RBD** | Reliability Block Diagram |

| | |
|---|---|
| **RBOT** | Rotating Bomb Oxidation Test |
| **RBS** | Resource Breakdown Structure |
| **RCA** | Root Cause Analysis |
| **RCFA** | Root Cause Failure Analysis |
| **RCM** | Reliability Centered Maintenance |
| **RF** | Radio Frequency |
| **RFP** | Request For Proposal |
| **RFQ** | Request For Quotation |
| **RMS** | Root Mean Square |
| **ROA** | Return On Assets |
| **ROCE** | Return On Capital Employed |
| **ROE** | Return On Equity |
| **ROI** | Return On Investment |
| **ROM** | Read-Only Memory |
| **RONA** | Return On Net Assets |
| **RPM** | Revolutions Per Minute |
| **RPN** | Risk Priority Number |
| **RTG** | Resistance To Ground |

| | |
|---|---|
| **RTY** | Rolled Throughput Yield |
| **RVP** | Reid Vapor Pressure |
| **SAE** | Society of Automotive Engineers |
| **SAS** | Shaft Alignment Specialist |
| **SCADA** | Supervisory Control And Data Acquisition |
| **SCM** | Supply Chain Management |
| **SDO** | Standards Development Organization |
| **SDT** | Self-Directed Team |
| **SDWT** | Self-Directed Work Team |
| **SI** | Système International |
| **SIC** | Standard Industrial Classification |
| **SIPOC** | Suppliers, Inputs, Process, Outputs, and Customers |
| **SKU** | Stock Keeping Unit |
| **SLO** | Standard Learning Objective |
| **SME** | Society of Manufacturing Engineers |
| **SME** | Subject Matter Expert |
| **SMED** | Single Minute Exchange Of Dies |
| **SMRP** | Society for Maintenance & Reliability Professionals |

| | |
|---|---|
| **SNR** | Signal-to-Noise Ratio |
| **SOC** | Standard Operating Conditions |
| **SOP** | Standard Operating Procedures |
| **SOX** | Sarbanes-Oxley Act |
| **SPA** | Shock Pulse Analysis |
| **SPC** | Statistical Process Control |
| **SQA** | Software Quality Assurance |
| **SQC** | Statistical Quality Control |
| **SQL** | Structured Query Language |
| **STEP** | Standard for The Exchange of Product model data |
| **STF** | Stress To Failure |
| **STLE** | Society of Tribologist and Lubrication Engineers |
| **STPD** | Standard Temperature and Pressure, Dry |
| **SUV** | Saybolt Universal Viscosity |
| **SWOT** | Strengths, Weaknesses, Opportunities, and Threats |
| **TAN** | Total Acid Number |
| **TAPPI** | Technical Association of the Pulp and Paper Industry |
| **TBD** | To Be Determined |

| | |
|---|---|
| **TBN** | Total Base Number |
| **TCE** | TriChloroethylenE |
| **TCP/IP** | Terminal Control Protocol/Internet Protocol |
| **TDR** | Time Domain Reflectometry |
| **TEEP** | Total Effective Equipment Performance |
| **THD** | Total Harmonic Distortion |
| **THERP** | Technique for Human Error Rate Prediction |
| **TLC** | Tender Loving Care |
| **TLC** | Tighten, Lubricate and Clean |
| **TOC** | Theory Of Constraints |
| **TPM** | Total Productive Maintenance |
| **TQC** | Total Quality Control |
| **TQE** | Total Quality Engineering |
| **TQM** | Total Quality Management |
| **Tr** | Transmissibility |
| **TRIZ** | TheoRy of Inventive problem Solving |
| **TTF** | Time To failure |
| **UCL** | Upper Control Limit |

| | |
|---|---|
| **UL** | Underwriters Laboratories |
| **UPC** | Universal Product Code |
| **UPS** | Uninterruptible Power Supply |
| **URL** | Universal Resource Locator |
| **UT** | Ultrasonic Testing |
| **UV** | UltraViolet |
| **VE** | Value Engineering |
| **VFD** | Variable Frequency Drive |
| **VGA** | Video Graphics Array |
| **VI** | Viscosity Index |
| **VOC** | Voice Of the Customer |
| **WAN** | Wide Area Network |
| **WBS** | Work Breakdown Structure |
| **WIIFM** | What's In It For Me |
| **WIP** | Work In Process |
| **WO** | Work Order |
| **WR** | Work Request |
| **WWW** | World Wide Web |

**WYSIWYG**   What You See Is What You Get

**YTD**   Year To Date

# Symbols & Numbers

**1.5 Sigma Shifts and Drifts**

The theory that, over time, any process in control will shift from its target by a value of up to 1.5 sigmas. Allowing for the 1.5 sigma shift results in the generally accepted 6 sigma value of 3.4 defects per million opportunities. Ignoring the 1.5 sigma shift results in a 6 sigma value of 2 defects per billion opportunities.

**3P**

The Product Preparation Process (3P) is a production preparation process tool used to support lean manufacturing environments. It is a highly disciplined, standardized, model. 3P results in the development of an improved production process where low waste levels are achieved with low capital cost. Synonymous with Pre-Production Planning.

**5S**

A lean tool used for workplace organization and standardization to reduce waste and optimize productivity. 5S drives a cleaner environment and organizes the workplace. The 5S concepts originated in Japan. They are:

Sort (Seiri)Set in Order (Seiton)Shine (Seis)Standardize (Seiketsu)Sustain (Shitsuke).

A sixth S (Safety) is sometimes added, subsequently changing the name to 5S plus or 6S.

**5W2H Method**

A methodology used to examine and question a process or a problem for developing improvement ideas. It engages a team in discovering overlooked issues or causes. It helps identify potential problems and develop optimum solutions by exploring the following questions:

What? – The Subject

Why? – The Purpose

Where? – The Location

When? – The Timing or Sequence

Who? – The People Involved

How? – The Method

How Much? – The Cost or Impact

**80/20 Rule**      A term referring to the Pareto principle. The principle suggests most effects come from relatively few causes. That is, 20% of the possible causes are responsible for 80% of the effects.

# Resources on the Web

### General Terms and Definitions
| | |
|---|---|
| Wikipedia | www.wikipedia.org |
| Business Dictionary | www.businessdictionary.com |

### Maintenance and Reliability
| | |
|---|---|
| ReliabilityWeb.com | www.reliabilityweb.com |
| Society for Maintenance & Reliability Professionals | www.smrp.org |
| Reliability & Maintainability Symposium | www.rams.org |

### Quality Management
| | |
|---|---|
| American Society for Quality | www.asq.org |

### Project Management
| | |
|---|---|
| Project Management Institute | www.pmi.org |

### US Government
| | |
|---|---|
| Occupational Health & Safety Administration | www.osha.gov |
| Environmental Protection Agency | www.epa.gov |

### Standards
| | |
|---|---|
| International Organization for Standardization | www.iso.org |
| American National Standards Institute | www.ansi.org |
| Canadian Standards Association | www.csa.ca |
| European Committee for Standardization | www.cen.eu |

# Bibliography

## Publications

Bloch, Heinz and Geitner, Fred, *Maximizing Machinery Uptime*, Elsevier, 2006

Campbell, John D. and Reyes-Picknell, James V., *Uptime, 2nd Ed.*, Productivity Press, 2006

Dalkir, Kimiz, *Knowledge Management in Theory and Practice*, Elsevier, 2004

Gulati, Ramesh, *Maintenance and Reliability Best Practices*, Industrial Press, 2009

Humphreys, Kenneth K. (ed), *Project and Cost Engineers' Handbook, 2nd Ed.*, American Association of Cost Engineers, Marcel Dekker, 1984

JIPM, *TPM Encyclopedia (Expanded Edition) Keyword Book*, Japan Institute of Plant Maintenance. 2002

Livingstone, John Leslie, *The Portable MBA in Finance and Accounting*, John Wiley & Sons, 1996

Martin, James William, *Operational Excellence*, Auerbach Publications, 2008

McKenna, Ted and Oliverson, Ray, *Glossary of Reliability and Maintenance Terms*, Gulf Publishing Company, 1997

Mitchell, John, *Physical Asset Management Handbook, 3rd Ed.*, Clarion Technical Publishers, 2002

Microsoft, *Microsoft Computer Dictionary, 5th Ed.*, Microsoft Press, 2002

MIL-STD 712C, *Definition of Terms for Reliability and Maintainability*, US DoD, 1981

Murno, Roderick A, et. al., *The Certified Six Sigma Green Belt Handbook*, ASQ Quality Press, 2008

Naryan, V., *Effective Maintenance Management: Risk and Reliability Strategies for Optimizing Performance*, Industrial Press, 2004

Nicholas, Jack Jr. and Young, R. Keith, *Advancing Maintenance, 2nd Ed.*, MQS, 1999

Palmer, Doc, *Maintenance Planning and Scheduling Handbook*, McGraw-Hill, 1999

Reliability Analysis Center, *Practical Applications of Reliability-Centered Maintenance*, US DoD, Alion Science and Technology, 2003

Shirose, Kunio (ed.), *TPM New Implementation Program in Fabrication and Assembly Industries*, Japan Institute of Plant Maintenance, 1996

Smith, Ricky and Hawkins, Bruce, *Lean Maintenance*, Elsevier, 2004

SMRP, *SMRP Best Practices Metrics and Benchmarking Glossary*, Society for Maintenance and Reliability Professionals, 2009

Stamatis, D.H., *Failure Mode Effect Analysis: FMEA from Theory to Execution*, 2nd Ed., ASQ Quality Press, 2003

Tague, Nancy R., *The Quality Toolbox, 2nd Ed.*, ASQ Quality Press, 2005

Tomlingson, Paul D., *Effective Maintenance*, Van Nostrand, 1993

Turner, Suzanne, *Tools for Success*, McGraw-Hill, 2002

Ward, J. LeRoy, *Project Management Terms – A Working Glossary*, ESI International, 2000

Watson, Gregory H., *Design for Six Sigma*, Goal QPC, 2005

## Papers

Kahn, Jerry D., *Applying Six Sigma to Plant Maintenance Improvement Programs*, MARCON 2006

## Standards

ANSI Z94.0-200, *Industrial Engineering Terminology*, Institute of Industrial Engineers, 2000

ANSI/PMI 99-001-2004, *A Guide to the Project Management Body of Knowledge, 3rd Ed.*, Project Management Institute, 2004

EN 13306:2001, *Maintenance Terminology*, CEN, 2001

IEC 60500, *International Electromechanical Vocabulary*

## Websites

ASQ Six Sigma Forum, Glossary of Terms
http://www.asq.org/sixsigma/quality-information/resources-sixsigma.html

Assembly Magazine, Glossary of Lean Manufacturing Terms
http://www.assemblymag.com

Equipment Reliability Institute (Wayne Tustin), Vibration & Shock Glossary
http://www.equipment-reliability.com/glossary.html

NIST/SEMATECH, e-Handbook of Statistical Methods
http://www.itl.nist.gov/div898/handbook

Maintenance Resource Center, (Sandy Dunn) Articles on Maintenance Terminology
http://www.plant-maintenance.com/maintenance_articles_terminology.shtml

STLE, Glossary of Terms
http://www.stle.org/resources/glossary

# Biographies

**Ramesh Gulati, P.E., CMRP, CRE, F.IIE**

Ramesh Gulati is Asset Management and Reliability Planning Manager with Aerospace Testing Alliance at the Arnold Engineering Development Center (AEDC) in Tennessee. Ramesh is a Certified Maintenance & Reliability Professional (CMRP), Certified Reliability Engineer (CRE), and licensed Professional Engineer (PE). He holds BSME, MSIE degrees and an MBA. He has authored many papers in the maintenance, productivity and reliability area, and is the author of the highly successful book titled <u>Maintenance and Reliability Best Practices</u> by Industrial Press (2009). Ramesh is very active in professional societies and in supporting educational institutions engaged in reliability and best practices education. He has served as the chair of the Best Practices committee and director of the Body of Knowledge directorate with the Society for Maintenance & Reliability Professionals (SMRP). Currently, he is director of Certification with SMRP. He also serves as an advisory Board member of the University of Tennessee's Reliability and Maintenance Center (RMC), as a member of Tennessee State University's industrial cluster board, and is a Fellow of the Institute of Industrial Engineering (IIE).

**Jerry Kahn, P.E., CMRP, F.NSPE**

Jerry Kahn is Marketing Manager for Maintenance and Reliability Programs for the Oil & Gas Division of Siemens Energy, and has over 35 years experience working for international technology companies. Jerry is a licensed Professional Engineer (PE) and a Certified Maintenance and Reliability Professional (CMRP). He holds BS and MS degrees in Chemical Engineering from Michigan Technological University and an MBA from City University of Seattle. Jerry is a Past President of the Georgia Society of Professional Engineers (GSPE) and Fellow of the National Society of Professional Engineers (NSPE). He is the USA delegate to the Technical Committee on Maintenance (COPIMAN) of the Pan American Federation of Engineering Societies (UPADI), and is a member of the Society for Maintenance & Reliability Professionals (SMRP), where he serves on the Best Practices Committee.

**Robert Baldwin, CMRP**

Robert Baldwin, an award winning author and editor, is a charter member of the Society for Maintenance and Reliability Professionals (SMRP) and is a Certified Maintenance and Reliability Professional (CMRP). He was the founding editor of Maintenance Technology magazine. He is a 45-year veteran of the technical business press, serving as associate editor, managing editor, and editor of a number of technical magazines. He is a graduate of Kansas State University with BS degrees in Civil Engineering and Technical Journalism.